Management for Professionals

The Springer series *Management for Professionals* comprises high-level business and management books for executives. The authors are experienced business professionals and renowned professors who combine scientific background, best practice, and entrepreneurial vision to provide powerful insights into how to achieve business excellence.

More information about this series at http://www.springer.com/series/10101

Timothy Jung • Jeremy Dalton
Editors

XR Case Studies

Using Augmented Reality and Virtual Reality Technology in Business

Springer

Editors
Timothy Jung
Faculty of Business and Law
Manchester Metropolitan University
Manchester, United Kingdom

Jeremy Dalton
PricewaterhouseCoopers
London, United Kingdom

ISSN 2192-8096 ISSN 2192-810X (electronic)
Management for Professionals
ISBN 978-3-030-72783-3 ISBN 978-3-030-72781-9 (eBook)
https://doi.org/10.1007/978-3-030-72781-9

This Springer imprint is published by the registered company Springer Nature Switzerland AG.
The registered company address is: Gewerbestrasse 11, 6330 Cham, Switzerland

Preface

Extended Reality (XR) technologies including Augmented reality (AR) and virtual reality (VR) technologies have made a tremendous amount of progress over the last decade. VR headsets have become simpler, cheaper, and more intuitive, making it easier for business leaders to build a case for adopting the technology. Meanwhile, AR has proliferated in multiple forms from head-mounted displays to the smartphones we carry around in our pockets.

These forces have resulted in AR and VR gaining a tremendous amount of traction and press coverage. However, this has mostly been related to applications in the consumer world, particularly entertainment—we rarely hear about how these technologies have been applied in businesses successfully for many years.

We have experienced many of these business applications firsthand—in some cases, we were involved in designing, developing, and deploying them—and so we could see the value of AR and VR to industry growing. This was the start of why we thought a project like this would be important. We wanted to bring together a diverse set of examples of both AR and VR being applied in different parts of the world, across different industries, and even within organisations pursuing different primary objectives from profit to social good.

The result is this book which brings together 19 detailed case studies starting with an initial introduction to each company and the AR or VR project they were involved in. This is followed by a deeper dive into the project which covers feedback from end users and a view on where each solution might go in the future. Each case study is long enough to provide a sufficient amount of detail but short enough to consume quickly.

We hope you enjoy exploring and learning from the organisations and projects featured in this book.

Manchester, UK — Timothy Jung
London, UK — Jeremy Dalton

Contents

There's More to Murphy; Virtual Reality in the Construction Sector

Gareth Riding

Abstract

A leading engineering and construction company, J Murphy and Sons Ltd., successfully developed and implemented a virtual reality training module with the assistance of Clicks and Links. This system allows Murphy to deliver engaging, emotive and safety critical content in a manner that improves knowledge retention and drives positive behavioural change across the business.

Keywords

Virtual Reality · Training · Safety · Engineering · Construction · Innovation

1 Company Description

J Murphy and Sons Ltd. is a leading global, multi-disciplined engineering and construction company founded in 1951. Murphy is passionate about improving lives by delivering world-class infrastructure.

Operating in the United Kingdom, Ireland and Canada, Murphy provides a range of construction services to infrastructure sectors including rail, water, power and natural resources. Headquartered in London, Murphy also boast a number of related businesses; Construction, Development & Property Services, Pipeline Testing Services, Ground Engineering, Utility Connections, Murphy Plant, and Urban Realm.

Murphy directly employs more than 3000 engineers, professionals and skilled operatives around the world. The company has skilled expertise in delivering pipelines, process engineering, design, marine, tunnelling, fabrication, bridges and

G. Riding (✉)
J Murphy & Sons Ltd, London, UK
e-mail: garethriding@murphygroup.co.uk

T. Jung, J. Dalton (eds.), *XR Case Studies*, Management for Professionals,
https://doi.org/10.1007/978-3-030-72781-9_1

piling. Murphy also invest heavily in their substantial holding of plant, equipment and facilities. As an infrastructure specialist, Murphy is able to respond to the challenges of any project while making sure to provide clients with a consistently high quality service.

Murphy's self-delivery model gives clients the assurance that the company can provide a safe, reliable and cost-effective service. Murphy make significant investments in plant and people, so as to offer a highly-skilled, experienced workforce who use modern and well-maintained equipment. This means Murphy effectively offer a 'one stop shop' service for clients on complex projects and can be on site and ready to work at the drop of a hat.

Murphy is able to offer a speedy and valued response to clients' emergencies and is the contractor of choice for many when unforeseen problems arise. The company actively encourages employees to share their ideas wherever they can, whether they work on site or in an office. These ideas have been put into good use on projects, helping to make Murphy's services safer and more efficient.

Murphy's ability to deliver projects in this way helps build and maintain strong relationships with clients. Murphy staff work hard to develop excellent personal relationships, listening to and understanding clients' requirements. They and their customers are at the heart of everything Murphy do.

2 Project Summary

Alongside Clicks and Links, Murphy's expert Culture & Engagement team have developed and implemented a virtual reality training module to supplement the delivery of a mandatory safety-oriented behavioural training module.

3 Project Details

3.1 Challenge

In order for Murphy to guarantee they are delivering world class infrastructure for their clients, the company must ensure the workforce are fully trained and competent in the work they are undertaking. Some of that training and experience occurs on live construction sites where there are multiple hazards (Fig. 1).

Large scale construction projects are dynamic and complex, and the combination of a range of hazards and variables on site results in an environment that changes quickly. This fluctuation in hazards and risks can put the most experienced workers in harm's way, as factors like weather, location and nearby work activities can change rapidly and combine to pose a serious threat. For individuals that are inexperienced in this field, there is no "one size fits all" approach to hazard awareness and risk management. The existence, location, nature, severity and potential consequence of any hazard can change an almost infinite number of

Fig. 1 Example of a complex construction site; Plumstead Tunnel with Tunnel Boring Machine (TBM) on Crossrail Thames Tunnel Project

times daily, therefore a habitual and subconscious risk perception approach is paramount to ensuring the safety of construction workers.

3.2 Solution

The solution to delivering safety information about a high risk environment? A fully immersive digital twin of a standard construction project, programmed with common yet variable hazards that may be encountered when on site.

The VR programme puts participants through their paces in a mock trip to site where they are asked to spot potential hazards. After putting on the headset, participants are transported to Murphy's reception in Kentish Town, before heading out to site.

The production of this model and the hazards therein draw upon Murphy's well documented historical safety performance. By reviewing the most pertinent risks and hazards across the business throughout the years, Murphy has developed a model that reflects and seeks to counteract known areas of poor safety performance. This tool allows the team to deliver "hazard awareness" content to the workforce and train their ability to recognise and mitigate risks on site.

The output from the session will inform users on where they performed well, plus highlight areas of their personal risk perception that need improvement. The tool works on a fix/consequence basis, as hazards that are correctly addressed will revert from the "risky" state to a safe state. Conversely, hazards that are not addressed will demonstrate to the user the serious consequences of lack of intervention. For example, there is an excavator located dangerously close to high voltage overhead power lines – if the users fails to spot this, the Culture and Engagement team will bring them back to the point and it will show the excavator touching the lines with sparks and smoke emanating from the machine. Data gathered throughout this task

Fig. 2 Screenshot of virtual reality model including dynamic hazard spotting

can be used to give the business more information about what training is required and where (Fig. 2).

This solution was developed as an alternative to standard virtual reality training modules, as they operate on a multiple choice basis that prompts the user to select a response at predetermined point in the model—this is not representative of a construction site and would not provide Murphy's workforce with adequate training and preparation. Instead, Murphy opted for a more natural approach that would better test the risk perception of its workforce, incorporating dynamic and randomised scenes that emulate real life conditions.

The system is also used to give people in an office a better understanding of site environments and risks even if they've never been on site before—optimising workflows between different functions and teams.

3.3 Benefits

A reduction in staff on site results in a reduction in risk. By exposing the workforce to a lower level of risk, Murphy can keep on living true to the company's 'Never Harm' value and its 'Fit to Start' strategic driver; making sure everyone is comfortable and proficient in their ability to question and action on site risks. This work ensures that the entire workforce, both existing and new, is aligned to the strategic direction of the company (Fig. 3).

The immersive nature of the VR system means that the users are absorbing more information about site hazards, better preparing them for the real thing. It is a safe environment in which to experience and explore the hazardous world of construction, allowing Murphy to fully brief and train its workforce from the safety of an office, eliminating the risk of on-site exposure to hazards. In accordance with the

Fig. 3 Virtual Reality training system in use by Murphy workforce

challenges outlined in the UK Government's Sector Deal and the Industrial Strategy, Murphy is actively investing in its staff to improve productivity, upskill the workforce and guarantee them a higher earning power.

4 Feedback from End Users

Following a successful nationwide pilot programme, there is excitement within Murphy as the business is leveraging modern day technology to an even greater extent. Users of the system have remarked upon the realisation that the system has highlighted their "factory blindness". For some experienced staff, a lifetime of working in high risk environments has bred in a level of complacency. This is actively challenged, identified and rectified through the use of the system.

Derek Stringer, Rail Supervisor, said "I thought the interaction with the VR was spot on, a total game changer. It really sets us apart on our training days. It really gives you something different to think about, seeing issues in real time and trying to get an understanding of what we should and shouldn't see out onsite.

I think VR has a big future in our industry, I don't see a reason why you can't use it to promote all the good things we do to our clients and even use it as a walkthrough of how future projects will look when complete.

A definite positive tick in the box for me."

Over 50% of the users would not usually come into contact with such advanced technology—due to either job role, reluctance or in some cases fear. The feedback from the majority of users has been that the system provides a user friendly and truly memorable introduction to some of the most advanced hardware and software currently in use within the industry. Feedback has shown that the use of this system has effectively helped reduce the reluctance and fear to embrace new technology.

The introduction of this system has sparked a wave of creativity and innovative thinking across the diverse Murphy family. The VR system is novel for most, and engagement with it has inspired the majority of users to challenge their own working processes in terms of safety, sustainability and efficiency, in turn driving the business and the industry forwards.

Karen Hartley, Rail Team Organiser, said "A great piece of kit that give you a realistic experience of navigating the dangers on site without being in any real danger, and great fun to use!"

5 Future Outlook/Roadmap

In the future, this technology could be used by Murphy as a standard response to safety incidents across the business, and the wider industry. By supplying the workforce with cheap "cardboard" type units, Murphy will be able to distribute highly engaging and emotive content around safety alerts and site inductions effectively, easily recording engagement and view count.

It could also assist in incident reporting as teams will be able to explore mock-ups of incidents, allowing a thorough root cause analysis to be undertaken and appropriate mitigations implemented.

Now that Murphy is using advanced virtual reality models with dynamic construction sites, the scope for future opportunities is vast. There are a plethora of potential applications within the design and engineering functions, including collaborative remote design review meetings, construction sequencing, and planning and logistics. Many designers are now adapting their workflows to use the virtual reality systems as their main design tools, meaning that future projects could be built entirely from within a virtual model with a greater level of speed and quality.

There is also an opportunity for Murphy to begin to generate 3D content within other functions such as safety, procurement, environmental, quality, operations—all of which can exploit the system to access contextual and relevant content in a much more engaging way.

Murphy continues to develop its ability to capture, analyse and interrogate data from a variety of sources. Going forwards, more opportunities will arise to integrate this system into existing workflows to improve the way that Murphy staff interact with the data they generate.

6 Conclusion

This project represents a positive step for Murphy, and is an exciting move towards a safer industry. The solution deployed by Murphy in response to training challenges, and to make sure staff are always fit to start, paves the way for further work in the virtual reality realm, bringing with it exciting new prospects concerning design, engineering, planning, logistics—all of which could see a step change in culture, collaboration and capability as a result.

Murphy's purpose, to improve life by delivering world class infrastructure, has been achieved through the successful implementation of this system. The VR training module both improves and sustains life of the workforce, their families, and the communities that Murphy operate in through an increased hazard perception and mitigation capability.

World class people deliver world class infrastructure; by investing in and improving the capability of its workforce, Murphy ensures that construction teams are operating in the safest manner possible. This minimises accidents, incidents, stand-down time/waiting and programme delays, guaranteeing world class delivery every time.

Acknowledgements J Murphy and Sons Ltd. would like to thank the Clicks and Links team for their support in developing a system that allowed the team to realise their ideas in a valuable and meaningful way, using cutting edge technology and innovative methodology.

Orthopaedic Surgical TRaining with Virtual Reality: OySTeR-VR

Alastair Konarski, Mike Williams, Ronnie Davies, and Bibhas Roy

Abstract

Manchester University NHS Foundation Trust developed and completed the OySTeR-VR project, comprising of the recording and production of 360° virtual reality videos of orthopaedic surgical procedures. These were collated into a library of high quality educational films, enabling a fully immersive educational experience for the training and development of orthopaedic trainees.

Keywords

Virtual Reality · Orthopaedics · Surgery · Training · Innovation

1 Company Description

Manchester University NHS Foundation Trust is one of the largest acute care trusts in the UK, providing comprehensive care to a population of 750,000 over 9 hospitals. Our vision is to improve the health and quality of life of our diverse population by building an organisation that excels in quality, safety, patient experience, research, innovation and teaching.

The Trauma and Orthopaedic (T&O) department delivers care to patients with a variety of musculoskeletal conditions from major trauma to chronic osteoarthritis. Through a large multidisciplinary team, we provide care to these patients, and our services include both emergency and planned surgery.

The NHS is responsible not only for the healthcare of patients throughout the UK, but also the training of the next generation of healthcare professionals. The training of consultant trauma and orthopaedic surgeons takes a minimum of 10 years after

A. Konarski · M. Williams · R. Davies · B. Roy (✉)
Manchester University NHS Foundation Trust, Manchester, UK
e-mail: bibhas.roy@mft.nhs.uk

T. Jung, J. Dalton (eds.), *XR Case Studies*, Management for Professionals,
https://doi.org/10.1007/978-3-030-72781-9_2

graduation from medical school. Manchester University NHS Foundation Trust provides training to these doctors, at various stages of their progression, as they rotate through a programme around the North West of England.

2 Project Summary

In collaboration with Manchester Metropolitan University and Mativision, the OySTeR-VR project was established to create a video library of surgical techniques, for a number of orthopaedic procedures. During the project, surgery was filmed in 360° high definition, and via a production team an educational virtual reality video was created, allowing the trainee to be 'inside' the operating theatre.

3 Project Details

3.1 Challenge

The application of virtual reality (VR) in surgery has been established for over 20 years, with particular interest in the role it can play in surgical training (Coleman et al., 1994). Traditionally surgical training has taken a master-apprentice model to develop the skills required for operating, however, over the past two decades there has been change due to a number of factors affecting training (Torkington et al., 2000). Around the world there have been restrictions on the number of hours surgeons can work, increased focus on patient safety and drive towards a more consultant-led service (Torkington et al., 2000; Badash et al., 2016; McCloy & Stone, 2001), which have led to a need for more training outside the operating theatre, and the dramatic advance in technology has enabled the use of virtual reality and simulators to have an expanded role.

Randomised Controlled Trials have reported the benefits of improved operating room performance following training on VR simulators for many years (Seymour et al., 2002), and a recent meta-analysis of trials investigating the use of VR in surgical training reported that it can reduce the operative time and improve operative performance in laparoscopic surgery (Nagendran et al., 2013). VR use has been expanded in surgical training in a number of fields including neurosurgery, vascular surgery, facial reconstruction (Badash et al., 2016) and orthopaedics, with reviews having found benefits in training for knee and shoulder arthroscopy using VR (Bartlett et al., 2018).

Medical applications and internet resources are increasingly used in training and clinical practice, and platforms such as YouTube and VuMedi allow clinicians to access videos to better understand surgical procedures and techniques (Vosburgh et al., 2013). The near universal ownership of smartphones amongst healthcare professionals and increasing number of devices with native AR and VR functionality, has allowed mainstream adoption of VR. Surgical videos can now be recorded in 360°and the next step is to stream such immersive videos as part of surgical training.

Throughout the process, it is vital to maintain attention to patient safety and anonymity. Informed consent of both the patient and healthcare professionals involved must be carefully considered and recorded. As with any electronic patient record, clinicians have a duty of care to maintain and protect patient confidentiality. Healthcare professionals must abide by the code of practice for confidentiality (NHS, 2003), and regulatory authorities for hospitals and clinicians highlight the importance of security for electronically stored patient data and has extensive guidance on handling patient information (GMC, 2017). In most countries, there are legal requirements to protect personal data, such as the GDPR across Europe, and all parties involved should be familiar with these (General Data Protection Regulation, 2016).

3.2 Solution

We present a case study of our experiences in recording shoulder surgery, for the development of virtual reality educational videos, and how we utilise a standard operating procedure to regulate the process of preparing and filming in an operating theatre using a 360° camera and close-up camera.

1. Pre-production meeting:
   A pre-production meeting or meetings where relevant with everyone involved in the project.
2. Meeting objectives:
   - Project brief is outlined discussing what is going to be filmed
   - Ideas, questions and concerns will be established including camera-rigging ideas, maintaining sterility and patient confidentiality.
3. Filming environment assessment:
   Assess the operating theatre where filming will take place. Considerations include:
   - Patient safety must be at the forefront
   - What steps can be taken to ensure patient confidentiality
   - Consider camera-rigging options
4. Pre-surgery preparations and rehearsal:
   Long before shooting the first film, practice or rehearsal sessions must take place in the operating theatre where filming will take place. This will serve as a great opportunity to
   - Practice camera and equipment positioning.
   - Test camera lighting settings and controls.
   - Learn and practice how to use recording and downloading functions on arthroscopic or endoscopic stack systems.

We worked through multiple set ups and found the following was the best set up for recording in our case, which was an arthroscopic shoulder procedure (Figs. 1 and

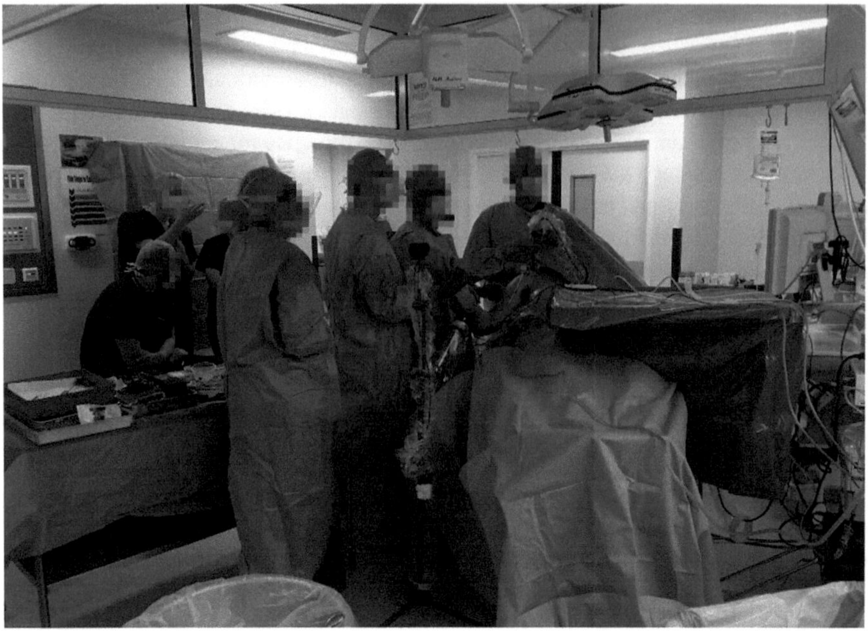

Fig. 1 Theatre set up for recording, with sterile drape for camera stand, and confidential information hidden

2). This will vary depending on the set up of the theatre and the operation being recorded. We found the best position for the 360° was level with the patient on the operating table, ensuring that the viewer using a VR headset can see this in a neutral head position, making viewing more comfortable. The camera stands were covered with a sterile drape, allowing positioning to a central location, without compromising sterility.

5. On the day of surgery:
   - Patient and actor consent forms must be completed before the patient comes to the operating theatre.
   - Be considerate and aware of sterility in the operating theatre.
   - Once set up is complete, an agreed upon signal (such as 2 loud claps) will identify the start of the recording, and allow the post-production team to sync the various recording devices.
6. Editing and voice over commentary:
   Post production editing is a key part of the final production and the clinicians should work closely with the film editing team to develop a clip that delivers the most educational value.
   - Films must be synchronized so that the 360° recording is compatible with the close-up views and overlays.
   - Post filming voice over commentary rather than live commentary.

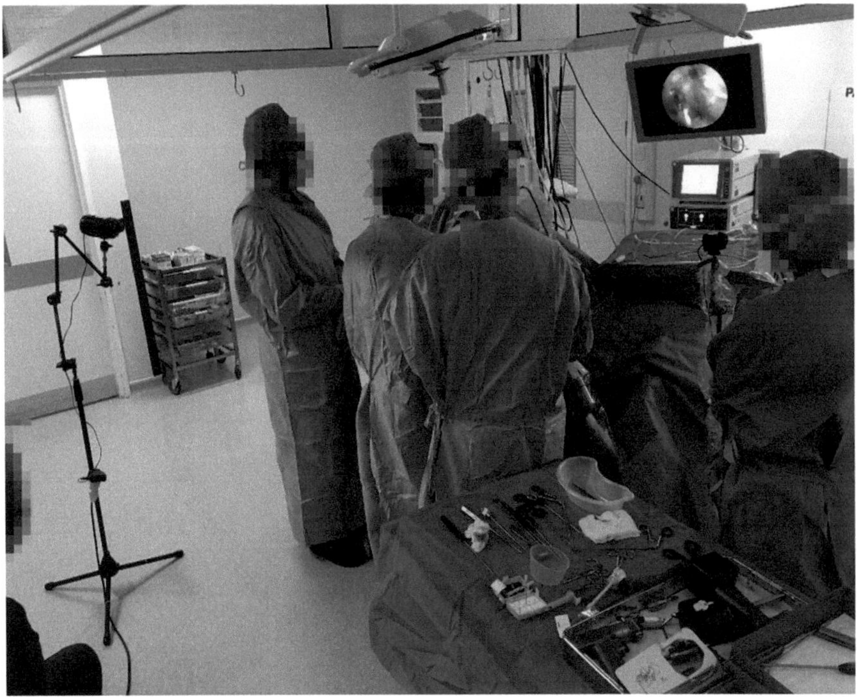

Fig. 2 Theatre set up for recording, showing position of close up camera

- The final production should highlight the steps in the surgery that are of most educational benefit, and we recommend limiting it to 15 min.
- The main focus during the filming can be overlaid and should be positioned centrally (Fig. 3). Accessory information, such as radiology images or anatomical diagrams should be available in the periphery.

3.3 Benefits

Recent changes in surgical training, and the way healthcare is delivered throughout the world have led to need for more training outside the operating theatre. The videos created allowed an immersive experience for trainees to see operations in a virtual reality setting. Through the video library, trainees were able to access these to supplement the training received in the operating theatre and the classroom.

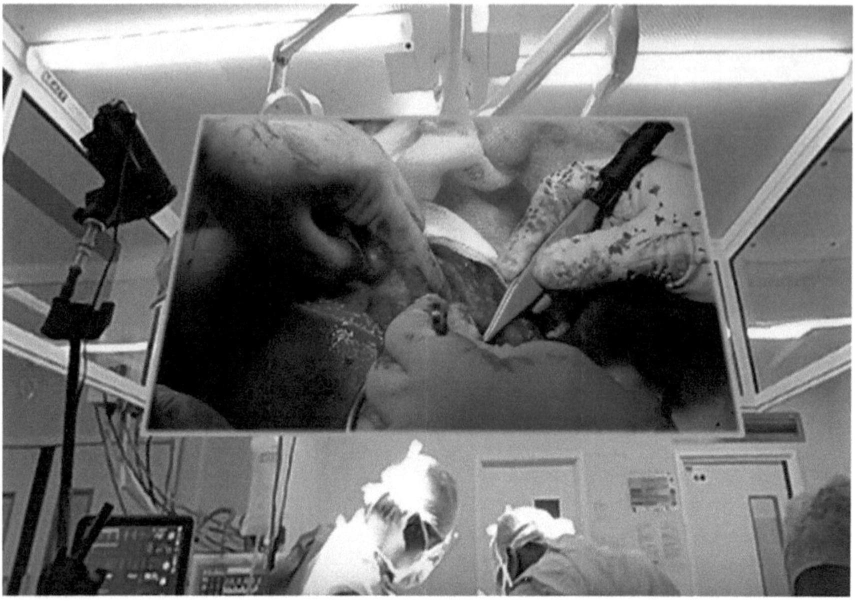

Fig. 3 Close up of surgery overlaid onto 360° film of operating theatre

4 Feedback from End Users

Following the production of the videos, trainees were interviewed on their experience with the virtual reality. Responses were very positive, and all participants enjoyed using the VR video. It was felt that this 360-degree video could be effective in providing a more detailed insight along with a clear explanation with the parallel PowerPoint presentation, when compared with a 2D video.

5 Future Outlook/Roadmap

There are multiple applications for VR in medical education, such as in the teaching of anatomy and virtual reality immersive recording of operations for training purposes (Khor et al., 2016). The training modalities outside of the operating room now include surgical simulators and multimedia presentations including surgical technique videos. Recent advances in VR technology and recording equipment has enabled the integration of surgical training videos with rich, interactive multimedia presentations in a truly immersive 360° virtual environment.

Further developments have included the first operation to be live streamed in 360° VR around the world in 2016 (Davis, 2016). This has since been replicated by many teams, and live surgery now forms part of many educational meetings (Khor et al., 2016). With the recent exponential increase in the role of VR in surgical training, there are now over 400 commercially available surgical simulators (Bartlett et al., 2018). It is estimated that the global augmented reality (AR) and virtual reality healthcare industry is already worth \$769 m and projected to be worth \$4.9b by 2023 (BusinessWire, 2018).

Manchester University NHS Foundation Trust, through its ongoing collaboration with Manchester Metropolitan University, is working on a number of projects to explore further uses of VR in healthcare. Such projects include the researching of VR and gamification in rehabilitation of musculoskeletal conditions, and further analysis of the use of VR in surgical training.

6 Conclusion

The OySTeR-VR project has been very successful in creating educational videos for surgical trainees. As the use and application of VR in medical education continues to grow, our case study demonstrates and provides a framework of how operations can be filmed in 360° in a safe manner, whilst protecting patient confidentiality and the laws governing data management, to create high quality immersive learning experiences.

The ever-improving technology combined with its increasing affordability means that the role of VR in surgical training will expand significantly over the next decade.

Acknowledgements Manchester University NHS Foundation Trust would like to thank Manchester Metropolitan University and Mativision Limited for their support in developing the educational videos through the OySTeR-VR project, along with the funding for the project from Innovate UK.

References

Badash, I., Burtt, K., Solorzano, C. A., & Carey, J. N. (2016). Innovations in surgery simulation: A review of past, current and future techniques. *Annals of Translational Medicine, 4*(23), 453.

Bartlett, J. D., Lawrence, J. E., Stewart, M. E., Nakano, N., & Khanduja, V. (2018). Does virtual reality simulation have a role in training trauma and orthopaedic surgeons? *Bone and Joint Journal, 100-B*(5), 559–565.

BusinessWire. (2018). Global Healthcare Augmented and Virtual Reality Market 2017–2023: Growing Need to Reduce the Healthcare Costs—ResearchAndMarkets.com 2018. Available from: https://www.businesswire.com/news/home/20180201005633/en/Global-Healthcare-Augmented-Virtual-Reality-Market-2017-2023.

Coleman, J., Nduka, C. C., & Darzi, A. (1994). Virtual reality and laparoscopic surgery. *The British Journal of Surgery, 81*(12), 1709–1711.

Davis, N. (2016) Cutting-edge theatre: World's first virtual reality operation goes live. *The Guardian*.

General Data Protection Regulation 2016. Available from: https://eur-lex.europa.eu/legal-content/EN/TXT/PDF/?uri=CELEX:32016R0679.

GMC. *Confidentiality: Good practice in handling patient information Manchester* 2017. Available from: https://www.gmc-uk.org/-/media/documents/confidentiality-good-practice-in-handling-patient-information%2D%2D-english-0417_pdf-70080105.pdf.

Khor, W. S., Baker, B., Amin, K., Chan, A., Patel, K., & Wong, J. (2016). Augmented and virtual reality in surgery-the digital surgical environment: applications, limitations and legal pitfalls. *Annals of Translational Medicine, 4*(23), 454.

McCloy, R., & Stone, R. (2001). Science, medicine, and the future. Virtual reality in surgery. *BMJ, 323*(7318), 912–915.

Nagendran, M., Gurusamy, K. S., Aggarwal, R., Loizidou, M., & Davidson, B. R. (2013). Virtual reality training for surgical trainees in laparoscopic surgery. *Cochrane Database of Systematic Reviews, 8*, CD006575.

NHS. *NHS Confidentiality Code of Practice* 2003. Available from: https://assets.publishing.service.gov.uk/government/uploads/system/uploads/attachment_data/file/200146/Confidentiality_-_NHS_Code_of_Practice.pdf.

Seymour, N. E., Gallagher, A. G., Roman, S. A., O'Brien, M. K., Bansal, V. K., Andersen, D. K., et al. (2002). Virtual reality training improves operating room performance: Results of a randomized, double-blinded study. *Annals of Surgery, 236*(4), 458–463. discussion 63-4.

Torkington, J., Smith, S. G., Rees, B. I., & Darzi, A. (2000). The role of simulation in surgical training. *Annals of the Royal College of Surgeons of England, 82*(2), 88–94.

Vosburgh, K. G., Golby, A., & Pieper, S. D. (2013). Surgery, virtual reality, and the future. *Stud Health Technol Inform, 184*, vii–xiii.

Innovation Flows Through VR for Water Utility

Michael Lewis

Abstract

Yorkshire Water commissioned a project with the AMRC to investigate the use of Virtual Reality (VR) during the design phase of their new 'standard product' catalogue. The project highlighted a number of design considerations which could be validated more easily, and rapidly through the exploitation of immersive VR, rather than the traditional approach of evaluating physical fluid handling prototypes which were very often scrapped. The use of VR not only produced significant cost savings for the company and consumers, it also enabled Yorkshire Water to reduce its carbon footprint as part of its mission to become a more sustainable utility through its Climate Change Strategy. The final phase of the project involved the successful implementation of a VR system which Yorkshire Water has incorporated into their product design lifecycle.

Keywords

Sustainability · Virtual reality · Ergonomics · Design

M. Lewis (✉)
The Advanced Manufacturing Research Centre (AMRC), The University of Sheffield, Sheffield, UK
e-mail: michael.lewis@sheffield.ac.uk

T. Jung, J. Dalton (eds.), *XR Case Studies*, Management for Professionals,
https://doi.org/10.1007/978-3-030-72781-9_3

1 Company Description

1.1 The University of Sheffield Advanced Manufacturing Research Centre

The AMRC, which is part of the High Value Manufacturing Catapult, is a network of world-leading research and innovation centres working with advanced manufacturing companies, large and small. The AMRC has a global reputation for helping companies overcome manufacturing challenges and has become a model for collaborative translational research involving universities, government and industry. AMRC research teams transform industrial and economic performance by making step changes in productivity, increasing competitiveness, developing new products and processes and training new talent and skills (Fig. 1). Their 110-plus industrial members range from global giants like Boeing, Rolls-Royce, BAE Systems and Airbus to small companies.

The AMRC's manufacturing expertise spans machining, advanced assembly and automation, augmented and virtual reality, robotics, casting, additive manufacturing, composites manufacturing, design, structural testing and training for the aerospace, automotive, energy, medical and other high-value manufacturing sectors.

More recently the AMRC has begun to apply this expertise to the construction sector, where advanced manufacturing techniques are now being seen as essential to increasing capacity, overcoming skill shortages and shortening lead times.

1.2 Yorkshire Water

Yorkshire Water is a private limited water supply and treatment utility company who provides water and sanitation services in Yorkshire and parts of North Lincolnshire and Derbyshire. As of December 2018 Yorkshire Water was providing water and

Fig. 1 The AMRC's Factory 2050 facility, an operator demonstrates advanced assembly technologies

sanitation services to over five million people including both domestic and commercial customers and was operating at a profit of £281.1 million (Yorkshire Water, 2018).

The following case study is a description of work undertaken by the AMRC that was funded by Yorkshire Water.

2 Project Summary

In an attempt to lower whole life costs of assets, improve delivery times, lower emissions and improve exports, Yorkshire Water began developing capabilities in standard product offerings (Browne et al., n.d.). These standard products would consist of modular fluid-handling equipment which could be assembled off-site, delivered and then fitted to larger fluid-handling projects. The development cycle required to generate a new standard product offering required Yorkshire Water to produce physical prototype mock-ups of standard product offerings to validate certain design requirements; however, it was deemed to be unnecessarily costly and did not provide operational teams with sufficient insight until relatively late in the development of the product.

With assistance from the AMRC, Yorkshire Water began evaluating virtual reality (VR) technologies for the design review process. The AMRC conducted a trade study of available software and hardware platforms which was presented to the standard products team at Yorkshire Water. Following a review, the HTC Vive head-mounted display (HMD) ("VIVE™|Discover Virtual Reality Beyond Imagination," n.d.) was selected as a suitable hardware offering and Virtalis' Visionary Render ("Visionary Render and Virtalis," n.d.) as a suitable software platform (Fig. 2).

The AMRC's developers then created a number of VR models, using Yorkshire Waters own standard product computer aided design (CAD) data, with the chosen software platform. This allowed Yorkshire Water to gain an insight into the VR content generation and the technologies required to realise their high level 'visualisation pipeline' (Schroeder et al., 2006). These were then reviewed by Yorkshire Water's design team and it was concluded that the VR models for the purpose of design review and validation was to replace their existing process, subsequently resulting in significant cost and time savings as well as reducing their carbon footprint. ("AMRC—Yorkshire Water—Irton Digital Twin—YouTube," n.d.)

3 Project Details

The accumulation of prototyping costs led Yorkshire Water to seek out alternatives to their existing design processes which involved building physical prototype mock-ups, each costing, on average, £40,000. These mock-ups were reviewed by design and operational teams to either validate their worthiness to enter production or to continue their design. As well as incurring significant prototyping costs, the standard

Fig. 2 A Yorkshire Water engineer reviews a water treatment tank in an immersive VR environment

products team had reported a number of occurrences where the design review process had failed. For example, it did not highlight issues between the interface between the standard product and the existing infrastructure to which it would eventually be fitted.

Yorkshire Water were aware of VR as a possible solution to improve design efficiency but were unsure how to apply the technology. It was at this point they made contact with the AMRC regarding their VR capabilities. The AMRC's digital team began by introducing VR concepts and technologies through case studies and demonstrations based on previously completed bodies of work. In addition, The AMRC also disseminated VR concepts by clarifying terminology and creating an unbiased basis to Yorkshire Water's understand of the technology (Fig. 3).

In the first phase of the project the AMRC worked with Yorkshire Water to gain an understanding of their design process, specifically, the software packages that were employed for standard product CAD and the modelling techniques used to create them. It was widely understood amongst VR practitioners that content generation, in this case the conversion of CAD models into a suitable format for viewing within a VR environment (referred to as the visualisation pipeline), is often the most challenging aspect of VR workflows. For that reason it was important that the AMRC had intimate knowledge of Yorkshire Water's design practices.

In the following phase, the AMRC were able to make recommendations for suitable software platforms on which to base Yorkshire Water's VR prototyping. A trial version of Virtalis' Visionary Render was selected as the software package due to its ability to handle large datasets whilst maintaining a high framerate. It was also selected due to its comprehensive list of CAD model importers. Yorkshire Water had identified that the most widely used CAD formats were Auto CAD,

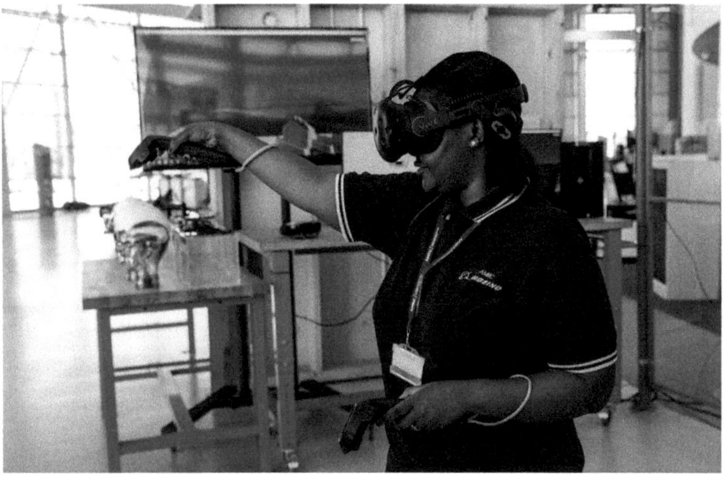

Fig. 3 The AMRC's purpose-built cell is used to demonstrate advanced visualisation technologies. A developer demonstrates the HTC Vive

Autodesk Revit, Autodesk NavisWorks and Dassault Systèmes SolidWorks although, due to their expansive supply chain, it could not be ruled out that additional model formats would be used in future.

For hardware, Yorkshire Water indicated a preference for a portable solution that could be taken between sites such that operational teams could be more readily included in the design phase of the standard product. For that reason, Virtalis' ActiveMove CVR ("ActiveMove CVR and Virtalis," n.d.) system was selected which included a HTC Vive HMD. The system was a complete 'turn-key' solution which is contained within a hard plastic case for easy transportation between sites (Fig. 4).

Following the hardware and software down-selection, the he AMRC project team created several VR prototypes using Yorkshire Water CAD data (Fig. 5). This phase sought to complete two major objectives. The first being the effective dissemination of the creation of VR prototypes to the Yorkshire Water design team using their own CAD data. This was achieved through practical workshops and technical demonstrations. The second objective was to effectively stress-test the VR system; for this activity, Yorkshire Water provided the AMRC with a number of highly detailed and complex CAD models such that creation of VR prototypes could be validated for a 'worst case scenario'. It was typical for a highly complex VR prototype to contain up to 14 million polygons and, to maintain useable frame rates, demanded advanced features within the Visionary Render software to be used such as, occlusion culling, level of detail and geometry instancing. The use of these techniques was, in turn, passed on to Yorkshire Water in an attempt to prepare them for future complex VR prototypes.

Finally, operational and design teams were invited to conduct VR prototype design reviews at the AMRC's advanced visualisation cell, located at Factory

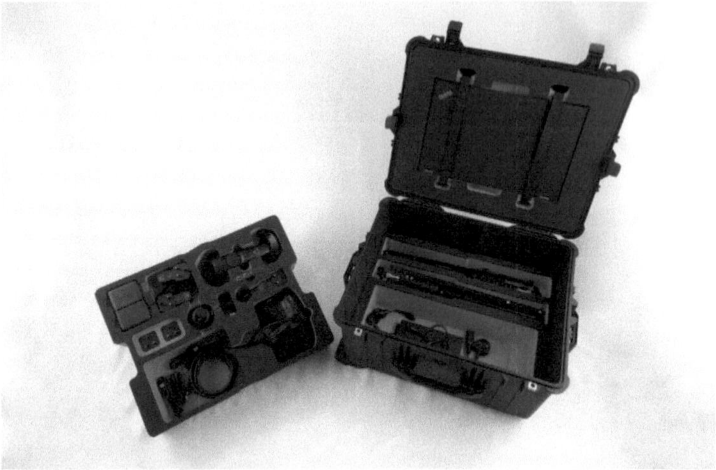

Fig. 4 Virtalis ActiveMove CVR a portable immersive virtual reality system

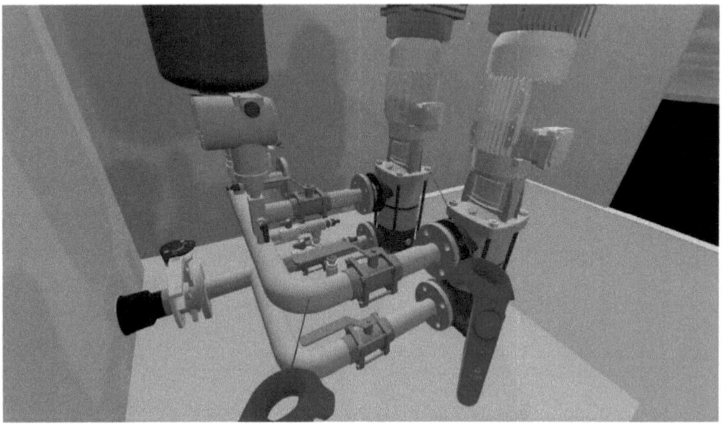

Fig. 5 CAD data was imported into the VR scene to evaluate the ergonomic implications of the design. The operative uses the hand controllers to assess reach and distance

2050 and on several of Yorkshire Water's project sites around South Yorkshire. The objective of the design reviews was to gauge the usability of the system with engineers and technicians who had no previous experience with VR systems. The result was a resounding success with an overwhelming majority of participants indicating a strong preference for the VR system when compared to their existing processes. Additionally, a significant number of operational staff were able to provide useful design input at a point in the development cycle that they would otherwise have never been involved.

As of January 2019, Yorkshire Water had invested in their own VR capabilities and continue to conduct design reviews for new products using VR as a core design tool.

4 Feedback from End Users

Over the duration of the project, Yorkshire Water had reported that it had saved £180,000 in physical prototype build costs and had projected that the savings would increase to £one million by the year 2020. It is also likely that these estimates are conservative as these costs to not consider the reduction of issues at the interface between new standard products and existing infrastructure.

Nevil Muncaster, Director of Asset Management at Yorkshire Water said: "What we now have is an immersive experience that enables us to check all the interfaces; to check that everything fits where it is supposed to fit before it goes on site; to ensure that safety and efficiency are fully integrated into the design, to give our operators and maintenance engineers a plant that is easy to run. It is a step change in how we design our new engineering projects and has the potential to generate significant cost savings."("Virtual reality gaming technology helping Yorkshire Water engineers design futuristic sites|AMRC," n.d.)

5 Future Outlook/Roadmap

Yorkshire Water are only at the start of their advanced visualisation journey. Following the developments outlined in this case study, they intend to stretch their capabilities further. The first stop on their roadmap is to evaluate technologies for connecting their virtual models with real-time data feeds to create digital twins of their assets. The intention is that this will allow them to make better and more informed decisions, faster. Further to this, Yorkshire Water would like to explore merging real-time data and simulated data with their VR models to create immersive training scenarios which could provide operational teams with valuable knowledge and crisis-management skills in flooding scenarios. The AMRC has also provided Yorkshire Water with an insight into augmented reality (AR) and its effective use for delivering digital work instructions and in-situ training manuals.

Yorkshire Water have created a number of 'digital champions' as a direct result of the work undertaken with the AMRC. These staff members are responsible for providing internal VR resources to staff members in an effort to promote its use in other areas such as health and safety and to ensure continued best practices in VR. They continue to engage with the AMRC on a regular basis.

6 Conclusion

Yorkshire Water were able to work with the AMRC to realise their new capabilities in VR for use in the design review phase of their development cycle. Through dissemination activities and scaffolding, The AMRC's visualisation experts were able to introduce Yorkshire water to new concepts and techniques. The AMRC hosted a number of workshops with Yorkshire Waters design engineers and operational teams to help them evaluate VR. At the conclusion of the project Yorkshire Water had adopted their own VR capabilities and subsequently reduced their design costs as a direct result of replacing physical mock-ups with virtual ones; they projected that savings of £one million could be made by 2020.

Acknowledgements Martin Rowe, Yorkshire Water.
Steve Forte, The University of Sheffield.
John Yates, The University of Sheffield.
Paul McColl, Virtalis.

References

ActiveMove CVR and Virtalis. (n.d.). Retrieved April 18, 2019, from https://www.virtalis.com/activemove-cvr/

AMRC—Yorkshire Water—Irton Digital Twin—YouTube. (n.d.). Retrieved April 18, 2019, from https://www.youtube.com/watch?time_continue=5&v=uf6C9dpFhhY

Browne, J., Taylor, L., & Wright, S. (n.d.). *The Water Hub A Standard Product Platform for the water industry Clients leading the way*. Retrieved from www..com.

Schroeder, W., Martin, K., & Lorensen, W. (2006). *The visualization toolkit, an object-oriented approach to 3D graphics*.

Virtual reality gaming technology helping Yorkshire Water engineers design futuristic sites | AMRC. (n.d.). Retrieved April 18, 2019, from https://www.amrc.co.uk/news/virtual-reality-gaming-technology-helping-yorkshire-water-engineers-design-futuristic-sites

Visionary Render and Virtalis. (n.d.). Retrieved April 18, 2019, from https://www.virtalis.com/products/visionary-render/

VIVE™ | Discover Virtual Reality Beyond Imagination. (n.d.). Retrieved April 18, 2019, from https://www.vive.com/uk/

Yorkshire Water. (2018). *Annual Performance Report 2017/2018*. Retrieved from www.yorkshirewater.com/reports

Making Training More Effective with Shared VR: Lanes Group Plc Improves Employee Retention and Reduces the Costs of Recruitment and Training

James Brewer and Jessica Davey

Abstract

Lanes Group plc uses an immersive projection cylinder as a collaborative way of engaging with virtual reality content. The cylinder, provided by Igloo Vision, is part of a wider programme of learning and development (L&D) and health and safety (H&S) initiatives. Overall, the aim was to reinforce the company's commitment to H&S, fast-track the induction of new recruits, and improve employee retention. In combination, the initiatives led to: a 57% reduction in employee attrition, a 9% reduction in employee feelings of unhappiness, and a £1 million reduction in recruitment and training costs. It therefore provides a good example of the efficacy of immersive content for training and development.

Keywords

Immersive projection · Shared VR · Virtual Reality · Training · Projection dome · Projection cylinder

1 Company Overview

This case study involves the work of three companies:

J. Brewer (✉)
Learning & Development Specialist, Lanes Group Plc, Leeds, UK
e-mail: james.brewer@lanesgroup.com

J. Davey
Igloo Vision Ltd, Wales, UK
e-mail: jessica.davey@igloovision.com

T. Jung, J. Dalton (eds.), *XR Case Studies*, Management for Professionals,
https://doi.org/10.1007/978-3-030-72781-9_4

1.1 Lanes Utilities

Lanes Group plc is the UK's largest independent drainage and utilities specialist.

One of its most prestigious contracts is a wastewater network service (WNS) maintenance agreement for Thames Water. Lanes Utilities, set up to deliver the contract, unblocks, cleans and repairs drains and sewers on behalf of 15 million Thames Water customers. It also delivers other specialist services, including wet well cleaning, tankering, and sewer rehabilitation. To keep the wastewater flowing across the 68,000-mile network, 880 field engineers and 220 support staff handle 1400 maintenance jobs a day.

1.2 Igloo Vision

Based in the UK, with offices in America and Australasia, Igloo Vision designs, develops and delivers immersive 360° projection domes, projection cylinders, and all of the enabling technologies. Describing itself a Shared VR company, Igloo Vision takes any VR or 360° content, puts it in a space that anyone can use, and makes it shareable among team members, stakeholders and clients. This means it becomes possible to work collaboratively (because everyone can see what each other is looking at), gauge reactions (because you can make eye contact and read facial expressions), and brief team members (because you can easily talk them through the VR content).

1.3 Myriad Global Media

Myriad Global Media is a digital communications agency specialising in innovative technology and content solutions. One of the company's specialisms is in designing and creating VR training programmes. The aim is to use immersive technology to make training more engaging, more effective, and less costly. And a particular focus is multi-user training solutions, especially for emergency response scenarios and health and safety initiatives.

2 Project Summary

An Igloo Shared VR cylinder was introduced by Lanes Group in 2017 as part of a wider programme of learning and development (L&D) and health and safety (H&S) initiatives. Overall, the aim was to reinforce the company's commitment to H&S, fast-track the induction of new recruits, and improve employee retention.

In combination, the initiatives led to:

- 57% reduction in employee attrition
- 9% reduction in employee feelings of unhappiness
- £1 million reduction in recruitment and training costs

Also, the company's training facilities have achieved a net promoter score (NPS) of 89%.

3 Project Details

3.1 The Situation

Lanes Utilities operates across the buoyant economy of South East of England, where the employment market can be fiercely competitive, especially for the type of roles that are vital to the company's Thames Water WNS contract:

- Field engineers—who can troubleshoot wastewater issues effectively and safely, apply the necessary remedies, and deal confidently and courteously with Thames Water customers and the general public.
- Scheduling and planning representatives—who can provide a top-quality response to Thames Water customers, many of whom may be facing a stressful situation (like sewer flooding to their property or sewerage leaks).

At the same time, the nature of the work had always presented some particular challenges to the company's learning and development team:

- A strong emphasis on health, safety and employee wellbeing—these are mission critical considerations, which need to be deeply embedded in the company's training programmes as well as its everyday working practices.
- A tricky employee induction dilemma—before they are allowed to enter potentially hazardous situations, newly-recruited field engineers need extensive training. Yet, until they have direct experience of such situations, it is almost impossible for them to appreciate the true nature of their new role. So, the induction programmes need to resolve this paradox.

The company was therefore looking for ways to improve and accelerate its induction programme, and bring more engagement to its health and safety initiatives—whilst reducing the overall cost of its recruitment and training programmes.

3.2 The Solution

In 2017, the company embarked on a full programme of inter-linked initiatives, including:

- A revamped approach to training—with an early emphasis on health and safety, plenty of re-fresher opportunities, a determination to invest in the latest technologies, and an aim to transfer the ethos of on-the-job training to a classroom environment.
- An innovative wellbeing programme—which emphasises the 'health' in health and safety, and includes a new mobile app, that requires operational staff to log how they are feeling before the start of their respective shifts.
- A new professional development programme—called Pathway to Success, to provide a fair and transparent pay structure, with full job checklists for every role, which enables employees to take more control of their own career progression.

As part of this wider programme of initiatives, Lanes convinced Thames Water to co-invest in an Igloo Shared VR system, in the form of a custom-built 7-metre Igloo Vision projection cylinder. Installed in its Customer Solutions Centre in Slough, this was to play a central role in the revamped approach to training.

3.3 Details of the Igloo Shared VR System

An Igloo Shared VR system is an all-in-one immersive technology solution, comprising of the 360° structure, the screens, the projectors, the servers, and the media player.

Systems come in various shapes and sizes, and range in capacity from teams of two-or-three people, right up to audiences of several hundred. The common denominator across all Igloo systems is the Igloo Media Player, the software that powers the system by:

- Ingesting any immersive VR or 360° content
- Distributing it via a series of ultra-high-definition channels
- Re-combining them into one seamless 360° edge-blended projection

Lanes wanted to accommodate groups of up to 15 people, so specified a 7-metre cylinder.

3.4 Why Lanes Was Attracted by Shared VR

Lanes had considered the use of immersive technology in its revamped approach to training, but was put off by the solitary nature of VR headsets. By contrast, a Shared VR approach presented several potential benefits:

- Sharing the experience
  Lanes wanted to retain the ethos of on-the-job training, but without the attendant risks and inefficiencies. In particular, the company wanted a way to combine

immersive technology with more traditional instructor-led training, enabling instructors to adapt sessions as necessary, and draw on the team dynamic, including:
  - Facilitating group discussions
  - Pointing-out potential hazards
  - Making eye contact
  - Reading body language
  - Seeing what everyone else is looking at
  - Arriving at a consensus opinion

  Lanes believed that, with Shared VR, it would be possible to immerse trainees in true-to-life scenarios, whilst still retaining the group dynamic.
- Incorporating field equipment
  As part of the induction process, Lanes needed to introduce new recruits to the type of equipment they would be using in the field, such as protective clothing, specialist tools, and safety monitoring equipment (such as hazardous gas detection units).

  Using Shared VR it would still be possible for participants to see, carry, and use the same equipment they would have with them in the field.
- Working with several types of immersive content
  Lanes could see an opportunity to work with a wide variety of immersive content, starting with basic films and, over time, incorporating more sophisticated content. For example:
  - Display standard 360° videos
  - Develop its own interactive 360° training content
  - Incorporate conventional presentation tools (like PowerPoint)
  - Explore the potential for the gamification of training The content-agnostic nature of the Igloo system, and its ability to integrate with game engine-based content, made all of this possible.

3.5 How Shared VR Was Integrated into the Training Programmes

Lanes established a long-term content strategy for Shared VR. The intention was to start out with simple 360° video (to get the team accustomed to the system), then begin to explore the wider potential of the technology and introduce more opportunities for interaction.

This has included:

- The "wow" video
  Lanes started by commissioning a high-impact 360° video, complete with spatial-surround sound, to introduce people to the scale and nature of the Thames Water wastewater operations, and some of the typical maintenance tasks. Intended

primarily for prospective field engineers, this is also a useful resource for scheduling and planning representatives (so they better understand the work of their colleagues), and other stakeholders (such as Thames Water representatives, and other visitors to the Lanes offices).

- The permutations

  Lanes invested in its own 360° camera and editing software. And its inhouse team has created around 40 different training scenarios. These are used for specific training modules, to give new recruits an in-depth introduction to the type of environments they will work in, the challenges they may face, and the remedies they will need to apply.

  These scenarios are also used for refresher training among more established employees. This helps Lanes to continually emphasise reinforce its commitment to Health and Safety, and to provide a mechanism for transferring skills and knowledge from highly experienced engineers to younger counterparts.

  With so many scenarios available to them, training leaders are able to assemble bespoke programmes, and also to link them together using conventional presentation tools (like PowerPoint).

- The gamification

  Most recently, Lanes has worked with Myriad Global Media to develop a fully-interactive mixed reality training application.

  Again, the aim is for an instructor-led session, and for a group of employees to participate. VR gameplay is introduced by using VR controllers. The application takes employees through a complete call-out scenario, and enables them to interact at each stage, including:

  - Initial customer call into contact centre agents
  - Assigning jobs to field operations employees
  - Carrying out vehicle and equipment safety checks
  - Complete tasks safely and successfully to solve a sewer flooding issue
  - Report back to a technical specialist to close the job The system requires each user to log in, entering their employee identification number, so all actions can be tracked via background analytics. This enables Lanes to track the progress of individual employees and highlight any areas of training that people struggle with.

3.6 Evaluating the Impact

The programme of initiatives enjoyed considerable success. And the Lanes team believes that the approach to training, with the Shared VR component as its centrepiece, made a strong contribution.

Highlights include:

- 57% reduction in employee attrition—a key consideration given the competitive employment market, which makes a significant contribution to the company's overall operational efficiency and effectiveness.
- 9% reduction in employee feelings of unhappiness—employees are actively encouraged to seek support that may improve their feelings of wellbeing, including the option of further training.
- £1 million reduction in overall recruitment and training costs—a significant bottom-line saving that more than compensates for the relatively modest costs of the Shared VR system and content

Also, on each visit to the training centre, employees are asked to evaluate their experience, leading to a net promoter score (NPS) of 89%.

With the introduction of the gamification initiatives, Lanes is also considering how best to formally evaluate the respective modules and calculate the return on investment—in order to provide a strong business rationale for the creation of further modules.

4 Feedback from End Users

The Shared VR system is deemed a definite success for Lanes. It is thought to have contributed to the improvement in employee retention, and the reduction in overall training and recruitment costs. And, as the centre-piece of the company's training centre, it has a strong impact on its net promoter score (of 89%).

To help evaluate the attitude of users and plan for future content, focus group research was also conducted. The response was very positive, as per the indicative comments below:

Question areas	Indicative responses
Overall impressions of the Igloo	"Overall, I think it's a good tool to have for our people and new starters. It's a different way of learning." "It doesn't fail to give me goose bumps when I go in the igloo. It's fascinating and exciting. It's a great opportunity to engage everyone from new starters to existing staff." "It's different, in a good way, from anything I have ever seen before in a company that I have worked with. It's fun, very interactive, and a new way of learning." "Outstanding technological achievement. It's user friendly and a game-changer."
Success of the Igloo in giving a feel for the type of environments Field Engineers work in	"It's hugely successful. Historically it's been hard to explain to Field Engineers what the day-to-day reality will be. The igloo gives a realistic feel to what the role will entail."

(continued)

Question areas	Indicative responses
	"It's very successful—as close to reality as you can get." "It's good preparation for the engineers. It's never going to be the same as real-life, but it does give a good insight into what it would be like. And I think it gives new engineers an air of excitement—rather than just listening to someone in a classroom or reading a book." "There is no substitute to the real thing, but this is as close as it gets—and provides a safe environment to do it in." "It gives us a different view that challenges assumptions. It also gives an insight into what the engineers need to do onsite."
Comparison with traditional training methods (like videos and PowerPoints)?	"It's interactive and immersive." "Far more engaging, exciting and interactive than videos and PowerPoint could ever be." "It doesn't really compare to be honest. It's completely different." "You can't get away from it—it's there in your face. It's loud, it's bright, and it grabs your attention straight way. Whereas with classroom-based learning you are easily distracted, which from a personal point of view isn't as fun." "It trumps them all, as it offers elements of all different types of training methods—you get the interactivity with the technology, as well as with the tutor, and the practical exercises."
The impression it gives regarding Lanes as a company—and its commitment to training	"That it will do what is ever necessary to ensure its people have the necessary tools and knowledge." "It makes it really clear that Lanes puts peoples' learning experience, training and development at the forefront." "It says Lanes are innovative and at the next level. Also, the investment in the technology and content shows the company cares about learning and development." "Lanes are committed to developing people. They're happy to spend the money if it means people will have more job satisfaction. It gives people an insight in what we do, and shows new recruits we care about their welfare." "Second to none. It shows the company is willing to invest in its people so they are well prepared before they start the role—and that also shows commitment to the client and their customers."

Meanwhile, feedback from leaders includes:

> The Igloo acts like a giant virtual reality headset. Up to 15 people can receive sewer maintenance training that's as life-like as possible, without them getting their feet wet.
> Andy Brierly, Director, Lanes Utilities.

> Traditionally, staff induction and training involved recruits doing a lot of sitting and being presented to. The Igloo breaks that mould. It brings wastewater training into the twenty-first Century, and is perfect for the millennial generation, brought up on gaming and digital TV.
> Mark Grimshaw, Head of Wastewater Networks North London, Thames Water.

5 Future Outlook/Roadmap

For the future, the aim is to continue to supplement the existing content, especially the new gamification content, by progressively adding in more scenarios and responding to emerging priorities (such as a new initiative to minimise the risks of pollution in watercourses, emergency response training, and new guidance from the Health & Safety Executive regarding the reduction of risk in temporary traffic management operations).

The long-term aim is to produce some 10–15 gamified training modules, and for them to be introduced at the rate of two-to-three a year. Also, the use of the system will be extended from awareness-building in the induction of new employees to competency-based assessments for more established employees.

The key point is that Shared VR will be deeply integrated into the way the company trains its employees, and that the content is continually added to and refreshed. This means that employees stay engaged, the team dynamic is always drawn-upon, and training managers are able to specify new content and address any learning gaps.

6 Conclusion

Although part of a wider programme of initiatives, it is thought that the use of Shared VR made a significant contribution to the overall success—and, particularly, the high Net Promoter Score (NPS) of the training centre.

From the Igloo Vision perspective there were five key factors that contributed to the success of the Shared VR deployment:

1.1. Success Factor #1: A solution to an existing problem

Delivering the right type of training to new recruits had always presented a dilemma to Lanes. Before they are allowed to enter potentially hazardous situations, team members need extensive training. Yet, until they have direct experience of such situations, it is almost impossible to appreciate the true nature of their new role. So, to get past this catch-22, Lanes wanted a way to immerse people in life-like scenarios as part of their induction.

By solving this dilemma, Shared VR delivered tangible value, and Lanes has benefitted from more effective training and improved retention of new recruits.

1.2. Success Factor #2: A commitment to training

When it first took delivery of its Shared VR system, Lanes put its key Igloo operators through an extensive training programme. So, from day-one, they understood the capabilities of the system and how to get the best out of it.

With a cadre of power users, Lanes has an on-site team of Shared VR experts, who act as rainmakers and trouble-shooters for the wider team.

1.3. Success Factor #3: A senior-level evangelist

From the outset, a member of the Lanes executive team (the Technical Director) was a strong advocate of Shared VR. In particular, he was keen to talk publicly and passionately about the business rationale for Shared VR and the benefits it could bring.

Senior-level support secured plenty of visibility for the Shared VR system via social media, traditional media, and internal communications channels.

1.4. Success Factor #4: Tight integration into wider processes and workflows

The Shared VR system is used as the centre-piece of the Lanes induction programme, which is mandatory for all new recruits. It is woven into the fabric of this programme, so is used on an everyday basis. And it has subsequently been used for on-going training requirements.

Shared VR is not an add-on. It is an intrinsic part of the way the company trains its people.

1.5. Success Factor #5: A long-term content strategy

At the outset, the company commissioned a basic yet high-impact 360° training film, to give new recruits a sense of what it is like to work within the wastewater network. From here, it plotted-out around 40 different interactive training programmes. It went on to develop complete VR training 'games'. And has a full roadmap of future enhancements.

The company has a disciplined approach to content creation. Starting with a basic film, and moving on incrementally, it has been able to apply its learnings. And, with a steady-stream of new content coming online, there is plenty to keep the teams interested and engaged.

Heroes to Meet in Virtual Reality; New Media Content Remediation

JongGil Song and JongSoo Kwon

Abstract

The Advance Media Tech Lab, established with the support of the government, created VR content using old photos, old media. Using 3D modelling or an actor requires a lot of money but using already created materials can produce VR content at low cost. Understanding the characteristics of VR and coming up with good ideas is the way to create effective VR content.

Keywords

Virtual reality · Remediation · New media · Old media · VR experience hall · Content production · VR movie · Seodaemun Prison · Advanced Media Tech Lab · Photography

1 Company Description

The Advance Media Tech Lab was established in 2017 with the support of the Korean Ministry of Science and ICT. Korea Economic TV, a leading Korean media company, participated as a cooperative company. Kyonggi University provided space and research personnel. The purpose of the organization was not to provide support for a small number of specialists, but to educate the public and students on new technologies, virtual reality.

The Advance Media Tech, located in the Seoul campus of Kyonggi University, is located in the heart of Seoul. Therefore, it is easy for the general public and students

J. Song (✉)
Department of Media & Visual Arts, Kyonggi University, Seoul, South Korea

J. Kwon
VRisVR Corp., Seoul, South Korea

T. Jung, J. Dalton (eds.), *XR Case Studies*, Management for Professionals,
https://doi.org/10.1007/978-3-030-72781-9_5

in Seoul to access it. These Korean government projects have three main purposes. The first is to build a facility capable of teaching virtual reality in a 300-meter-square space. We have created 10 booths with 10 virtual reality devices to achieve the purpose of educating many people, not facilities for a small number of researchers. One booth was built in a separate room with a size of 2.5 meters square. For the immersive feeling, the connecting line of the HMD was designed as special equipment and installed on the ceiling. Designed with rail and spring devices, the HMD assistive devices help users to experience virtual reality and move around. For safety, the monitor was installed outside the booth and the computer was put on the wall to clean the inside of the booth.

The second goal of the project was to provide virtual reality education to many people. To this end, we prepared the virtual reality contents and education curriculum necessary for youth and general education. Educated people are divided into students and ordinary people, and divided into people who experience once and those who participate in high-level courses. Cooperation with the local community was important to recruit participants. The Korean government operates a variety of programs for vocational education for young people, and the organizations responsible for the programs are located in each region. We collaborated with the organizations and made a lot of effort in the training of the vocational experience using virtual reality. In 2017, 2760 people provided virtual reality education, and by 2018, 9355 people were carrying out virtual reality education.

Finally, Advanced Media Tech Lab produced virtual reality contents. The purpose of the production of virtual reality contents was to promote the virtual reality industry and to give educational effect to the public. Among the various ideas at the planning stage, it was decided to create content for Seodaemun Prison, which is very close to Kyonggi University. Korea was ruled by Japan for 35 years (1910–1945) in the early 1900s. During that period, there are many heroes who have endeavoured for Korean independence. Seodaemun Prison is the place where many Korean independence fighters were kept under the control of Japan and is now used as a place to commemorate that time. Many students come to visit as part of their school classes. It preserves well the original form of the prison, and it shows the cruel torture of Japan and the suffering of the Koreans by making wax figures for the situation of that time. We chose one of the prisons to make a virtual reality content that meets the hero who was in that prison. It is important for young people to convey patriotism and history education in an interesting and entertaining way (Fig. 1).

2 Project Summary

Advanced Media Tech Lab produced virtual reality contents in Seodaemun Prison in Seoul. In order to produce popular and educational virtual reality contents on a limited budget, we used existing historical data and animation technology.

Fig. 1 The Advance Media Tech Lab

3 Project Details

Korean independent hero virtual reality contents began planning in September, 2017. The goal was to create content that allowed visitors to the Seodaemun prison to have greater impression and historical education. The content of the third part is divided into the first intro part, the actual hero's talking part and the last ending scene. The intro and ending were made possible to skip, and in the scenes where the heroes were talking, the user was able to choose between two stories. Experiences must listen and respond to the heroes. The reason for continuing this choice is for the active participation of the participants.

The sources of our virtual reality content are all historical records. It is a fact that comes from the photographs taken by the Korean independent heroes while they were alive, the letters they exchanged, and various records. Therefore, virtual reality viewers will see the most-watched hero's photos again through the HMD and listen to the most famous words left by the hero. Our goal was to bring existing historical data effectively into virtual reality. This method not only saves production costs but is also very efficient.

Finally, it was very important to distribute our virtual reality contents. We have released two versions that can be downloaded for free from Google Play Store and experience using Samsung Gear VR. This was the best choice at the end of 2017. However, the contents of the Google Play Store were not responded to because of the difficulty of advertisement and the 360 screen resolution. We operated an experiential zone using Samsung gear VR at Seodaemun prison. Since the virtual reality devices were available in small quantities, it was difficult for the group visitors to experience it. In order to improve these points, we set up an independent large space in 2018 and facilities for 15 people to experience together (Fig. 2).

Fig. 2 Seoul Seodaemun Prison and the prison inside the past scene reproduction dolls

3.1 Challenge

The budget allocated for the production of virtual reality content in 2017 was about $ 50,000. The researchers at the institute were paid separately, and they undertook the role of planning, scenario, supervision and producer. We made progress with computer graphics and unity work to external game companies. The amount of $ 50,000 was not enough to shoot a virtual reality movie, and it was not enough to make a good game. And most importantly, the content we were going to create was not a movie or a game. The solution to these difficulties was to borrow content made from other media. The content that was created in the past media is applied to the new virtual reality media.

3.2 Solution

Most of the virtual reality contents are created through 360-degree shooting or 3D modelling. But both are expensive and time-consuming tasks. So we studied how to reuse existing historical data to take on virtual reality media in order to make good content on a small budget. Instead of acting heroes or creating a virtual model of computer graphics, the heroes seemed to be moving and talking using stereoscopic and animated effects of well-known hero's photographs. Of course, some of the experience might be disappointed that there are no innovations, but it was possible to deliver a stable and realistic story rather than awkward acting or remote computer graphic characters (Fig. 3).

3.3 Benefits

People who first experience virtual reality through HMD become very interested. Virtual reality, which takes up the entire field of view, provides users with a great immersion experience. It is now a fact that fear or interest-oriented contents are the main thing in order to double the feeling of immersion and presence, or to emphasize

Fig. 3 A scene of Korean Independence Heroes Virtual Reality Content

differentiation from other media. Differentiation from existing media is important, but virtual reality media can be used more widely. It is good to have a dramatic experience in a short time, but it can be used more effectively when delivering boring and unpopular content. In Korea, history education is often excluded from priorities. Language, mathematics, and foreign languages play the most important role in the university entrance examination, and young people are reluctant to recall the days of historical hardship. Virtual reality contents are an effective medium to bring out stories that are more effective and familiar to these youths. Therefore, it is our hope that many educational virtual reality contents will be created.

4 Feedback from End Users

The responses of the participants were positive. Most of the virtual reality contents are produced as an excitement, and there are many opinions that it is scary or dizzy. However, most of the opinions of participants who have experienced Korean independence hero contents are very stable and educational. Young people were delighted to experience a vivid experience of meeting forgotten Korean heroes right in front of them, and the elderly were amazed to see living heroes. The stories of the heroes who devoted their lives for the independence of Korea became more and more touching to the experience.

We built a facility to watch VR in the space provided by Seodaemun Prison. The management of the equipment and the operation of the experience hall are handled by Seodaemun Prison. Seodaemun Prison had some difficulties that had to be solved when the VR experience facility was first operated. Firstly, training of operating personnel is difficult. The people who work at Seodaemun Prison are volunteers or

Fig. 4 Seodaemun Prison virtual reality experience hall

young men in military service, who are frequently replaced. Every time the working people change, new training of difficult VR equipment has to be done. Seodaemun Prison wants more manageable equipment and an intuitive content menu. It is said that it is best for the experienced person to enjoy the contents by themselves without any explanation. This may be our challenge as the biggest problem is the inferior mechanical completeness of HMD's appreciation of VR (Fig. 4).

5 Future Outlook/Roadmap

In the future, existing contents should be re-used for virtual reality media. 360-degree stereoscopic photography will be a large-scale project from the technical point of view as well as the limited costumes and props of the actors. The case of 3D modelling is similar. To create a complete 3D character, one designer must do more than a month of hard work. But virtual reality content does not have to be new. Rather than using the existing ones completely, transformation and redeployment are necessary, but they are much easier than creating new ones. Many of these cases have been found in the past. We can find many examples as voices are transferred to books, books are also transferred to radio and movies, and this is transferred back to computers and smartphones. Although there are many obstacles to dissemination and diffusion of virtual reality media, it is necessary to make many contents along with the development of hardware in order to solve this problem.

We produced VR contents for a specific place called Seodaemun Prison. However, the next project is preparing VR contents for more general students. Job experience content is that. One of the greatest features of VR's content is that it enables an inexplicable experience. It is possible to experience flying in the sky or getting into the water. And the advantage is that the experience is very realistic. So our goal is to enable students to experience something that is impossible in the classroom. For example, students will be given the task of building robots in the world of the future while giving them the programming and engineering training they need. As many students prefer to computer games rather than read books, job experience VR content will help them learn as much as they enjoy computer games. We also believe that virtual experiences of the future can give them dreams and hopes.

And we want to present such content as a mobile service. The problem with Seodaemun Prison is that users must come to a fixed space. On the contrary, the equipment is mounted on a mobile truck and moved to a place requiring education to provide a service. If a VR experience hall is installed in a particular school, only the students of that school can use it, but if they move around making it mobile, they can teach VR education whenever they need it. This can be seen as a form of sharing economy.

6 Conclusion

Audiobooks read the content written in an existing book. Here, the content changes from visual to auditory. But the contents remain intact and do not take much effort. In the past, books made of paper were inexpensive and easy to move, but nowadays many people have access to smartphones and audio devices, so audiobooks can be used without difficulty. Virtual reality media would also be the same. There are still problems with computer speed, screen resolution, and wireless transmission speed, but this will be solved gradually. Then, various contents will be transferred to virtual reality media. Virtual reality media is an intermediate stage to the hologram media, and even if a complete hologram is realized, it will continue to be used due to human physical conditions. It is natural for companies that adapt quickly to new media to achieve great success, and most companies are now investing heavily in virtual reality media. More new attempts should be made to get rid of vague fears about virtual reality media and abandon obsession with existing media.

Use of Augmented and Virtual Reality for Enhancement of Aerospace Cabin and Cockpit Experience

Maryam Safi and Joon Chung

Abstract

Ryerson University's Mixed-Reality Immersive Motion Simulation (MIMS) lab conducts a variety of research on Mixed Reality (MR) application for aerospace cockpits and cabins. The research team at the MIMS aims to integrate Commercially Available Off-the-Shelf (COTS) equipment with flight simulator data for faster and more reliable development of aerospace systems. The MIMS team has successfully used MR in crew training, rapid development of aircraft cockpits, and visualization of cabin interiors. This chapter provides an overview of three research projects completed at the MIMS lab using MR technology.

Keywords

Aerospace applications · Aircraft cabin · Aircraft flightdeck · Human-Machine Interfaces (HMI)

1 Company Description

The Mixed-Reality Immersive Motion Simulation (MIMS) lab is a research facility at Ryerson University's Department of Aerospace Engineering located in Toronto, Ontario, Canada. The MIMS lab is home to a wide variety of research initiatives including Mixed Reality (MR) simulation and aircraft interiors. MR is a term used to describe Augmented Reality (AR), Virtual Reality (VR), and their variations collectively (Safi et al., 2019). The lab is supervised by Dr. Joon Chung and is populated with Ryerson graduate and undergraduate students (MIMS, 2016). Software and

M. Safi · J. Chung (✉)
Department of Aerospace Engineering, Faculty of Engineering and Architectural Sciences, Ryerson University, Toronto, ON, Canada
e-mail: maryam.safi@ryerson.ca; j3chung@ryerson.ca

T. Jung, J. Dalton (eds.), *XR Case Studies*, Management for Professionals,
https://doi.org/10.1007/978-3-030-72781-9_6

hardware capabilities include fixed-base and full motion simulators, high fidelity flight simulation software, and numerous head-sets for AR and VR development. In 2018, the MIMS lab acquired a cabin mock-up of the now Airbus A220, which is used for cabin interiors research and development.

2 Project Summary

2.1 Augmented and Virtual Reality in Aerospace Applications

AR was first used in aviation as gun sights on board World War One fighter aircraft (Safi et al., 2019). Years later, the invention of the Head-Up Display (HUD) and Head Mounted Display (HMD) inspired researchers and engineers to apply AR tools in daily aerospace operations (Austakalnis, 2017; Frigo et al., 2016). Today, MR is used in various categories of the aerospace industry, including engineering support, navigation, and training and simulation. With the popularization of AR and VR, researchers are investigating the feasibility of using Commercially Available Off-the-Shelf (COTS) MR equipment in airport operations, crew support, and In-Flight Entertainment and Communication (IFEC) (Billinghurst et al., 2015; Morrison, 2018; Peddie, 2017).

2.2 MR Projects at the MIMS Lab

The MIMS lab works with academic and industry partners to improve the aerospace experience for passengers and crew. The lab focuses its MR research on two fields: cockpit development and cabin development. A primary goal for cockpit development is to integrate flight simulator data with COTS AR and VR equipment to allow for faster, reliable development of aerospace specific applications. Human-Machine Interfaces and Interactions (HMI^2) in the cockpit are studied to find optimal methods of using MR devices to alleviate pilot workload and provide safer operations. This field of research also looks for methods of using MR for pilot training, rapid flight deck development, and prototyping. The cabin research group investigates ways of using MR to enhance the cabin experience for passengers. Initial ideas include using MR for visualization of interior components, augmenting cabin windows with information, and providing IFEC using MR technology.

3 Project Details

3.1 Pilot Training Module Using Augmented Reality

One of the first AR projects at the MIMS lab is a pilot training module for take-off procedure. The module uses the Microsoft HoloLens to provide student pilots with multimedia instructions to aid in successful completion of a take-off manoeuvre using an Airbus A320 simulated cockpit. The application developed provides

Fig. 1 Left side: Ryerson Fixed-Based Simulator is used to test the pilot training module. Right Side: snapshot of the pilot's view when using the module with text and image instructions visible

student pilots with text, video, audio, and supplementary image instructions. The tutorial module is initiated when the pilot turns on the aircraft batteries and ends at the second phase of ascend.

The module is designed to take full advantage of the HoloLens' interactive ability, such as the use of hand gestures and voice commands to control the module. Once initiated, the tutorial displays instructions in a sequential order of steps. The module offers control options to return to the previous step, proceed to the next step, and start from the beginning. Only one step is displayed at a time and remains displayed until the user inputs a NEXT command using gesture or voice input. Figure 1 shows a screenshot of the pilot's view when using the tutorial. As seen in the figure, text information as well as images of the controls are provided to the user. Beyond the augmented text and image information, the simulator screens can be seen.

3.2 Interactive Virtual Aircraft Cockpit

A recently launched project at the MIMS lab uses VR headsets and motion sensors to recreate an interactive cockpit of the A220. Typically, user interactions with VR are done using controllers or haptic gloves, both of which are cumbersome to use and limited in their ability to simulate real HMI[2] in the cockpit. The introduction of motion sensors like the Microsoft Kinect and the Leap Motion Controller have allowed developers to track human movements for a wide variety of uses. Such sensors enable more natural interactions between user and machine as they can track hand motion and finger positions. Movements like button clicking, switch flipping, and finger wiggling are recognized easily using motion sensors and can be used as different input parameters to allow interactions with the virtual system.

The interactive virtual aircraft cockpit created at the MIMS lab uses a Leap Motion Controller mounted to an HTC Vive VR headset to track the user's hands in space. The VR headset displays the graphical representation of the aircraft cockpit, as shown in Fig. 2, which are viewable to the user once the headset is worn. Using Unity 3D environment, elements of the cockpit were made to be interactive. These include the throttle stick, landing gear lever, various switches, and scroll wheels. The Leap Motion Controller tracks the user's hand movement in space and recognizes interactions

Fig. 2 Snapshot of a user's view when wearing interacting with the virtual cockpit. The hand shown is tracked by the Leap Motion Controller

between the user's hands and the interactive virtual objects. The registered interactions are sent to the system as input parameters, which result in various output responses. The output responses are displayed on the VR headset by changing lever, switch, or scroll wheel position between ON/OFF states.

3.3 Head-Mounted Display Using 2D and 3D Views for Aircraft Flight Simulation

The MIMS lab has been interested in using the Microsoft HoloLens with COTS flight simulators to create a HMD. As a standalone computer system, the HoloLens can be used untethered, eliminating the user's need to remain within a specific radius of the operating computer. Additionally, the HoloLens can receive user input, creating an interactive HMD (Chittaluri, 2017).

The HoloLens' limitations on processing power and memory space largely constrain developers when working with the device. The graphical output format of the HoloLens is different from that used by COTS cockpit and HMI^2 development. Many industry standard HMI^2 software use OpenGL graphics, while the HoloLens uses DirectX11. Currently, there is no accessible or easy method of communicating graphical information between the two graphic formats.

In response to this challenge, the MIMS lab worked on a simple HMD for flight simulation that can receive simulation data, then present the information onto the HoloLens. The device communicates using Bluetooth Load Energy (BLT). Data is sent from the simulation software via BLT to the HoloLens. The HoloLens listens to the data, then draws out the graphical representation of the interface defined by the data received. As a result, the user can see real-time flight data on the HoloLens.

The team further expanded on the project by incorporating 2D navigation symbology and 3D view of the aircraft. The 2D navigation symbology mimics standard HMDs and flight navigation instruments. The 3D view showcases an aircraft with its

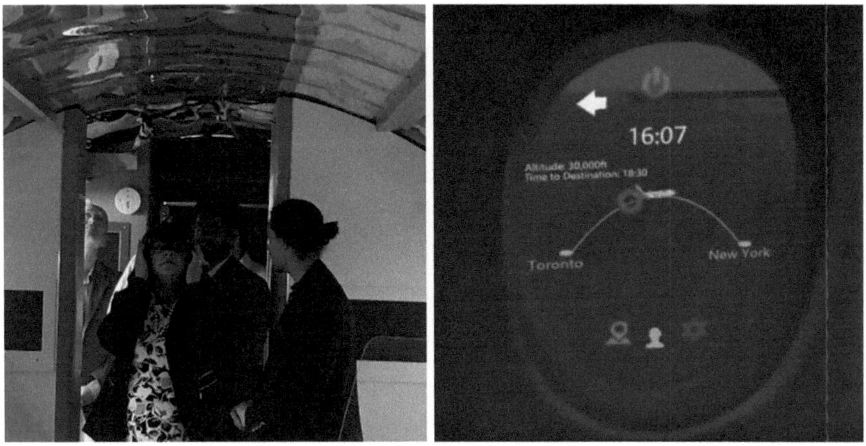

Fig. 3 Left Side: woman wearing the HoloLens explores the interiors of the A220 cabin. Right Side: AR demonstration is shown overlaying the cabin window

main axes labelled. The aircraft's position relative to the lateral, longitudinal, and normal axes is updated as information is sent from the simulation software to the HoloLens. Hence, if the pilot pitches down 45°, the 3D view would show an aircraft with its nose pointed 45° below the lateral axis.

3.4 Augmented Reality Technology for Visualization of Aircraft Interior Components

In cabin interiors development, the MIMS team successfully used AR headsets to prototype and demonstrate potential cabin improvements and installations. The concept of the AR window and the windowless aircraft are both being investigated at the lab, and demonstrations for both concepts are currently conducted using AR and VR headsets. The end goal is to use flexible OLED panels and transparent screens to augment aircraft windows and cabin dividers with information of interest or entertainment for passengers. The windowless aircraft concept would utilize these screens and panels for aircraft interior walls, enabling passengers to view content of interest along the walls of the cabin. Figure 3 shows the AR window demonstration developed by the MIMS lab.

4 Feedback from End Users

4.1 Pilot Training Module Using Augmented Reality

Some positive feedback was received from users of the HoloLens training module. Providing audio and visual information was received positively by pilots who tested the module. However, the placement of information was found to need improvement. Users reported that images and text were sometimes unclear clear or difficult to see, which hindered their performance and caused an increase in workload. In this case, the user was required to put in extra cognitive work to understand the visual information provided by the module, which went against the objectives of the project.

Users also responded favourably to the ability to control the pacing of the tutorial. Since the module does not proceed to the next step until a command is received, the pacing of each step was entirely up to the user. This enabled users to spend more time on steps they found difficult without needing to continuously replay or return to previous instructions. In this way, users felt less stressed and anxious, as the module did not proceed without their consent.

4.2 Interactive Virtual Aircraft Cockpit

The interactive virtual cockpit received a tremendous amount of positive reviews due to its ease of use and set up. The Leap Motion Controller enabled users to comfortably interact with the VR cockpit in a way that was natural and non-disruptive. Users were especially impressed with the realistic interactions. They were able to interact with VR switches, levers, and buttons in the same way they would with physical components. This was of interest to many corporation representatives as it provided an insight faster and cheaper cockpit prototyping and training opportunities.

4.3 Head-Mounted Display Using 2D and 3D Views for Aircraft Flight Simulation

The initial purpose of this project was to connect COTS simulation software to the Microsoft HoloLens. The 2D symbology and 3D view was added to show information flow between the software and the HoloLens. When presented, the project was received favourably. The information flow between a computer running the simulation software, and the HoloLens's on-board computer was of interest to many aerospace cockpit developers. This project enables the use of industry standard software to develop cockpit environments that can be tested with a remote, untethered AR device. Questions of clutter arising from the presentation of 2D and 3D views were raised by a few users. Since the purpose of this project was not to develop optimal symbology, this concern was not addressed at this time.

4.4 Augmented Reality Technology for Visualization of Aircraft Interior Components

As a visualization tool, this project received positive feedback. While the hardware performance was not perfect—the displays experienced latency or occlusion problems from time to time—users were satisfied with the ability to visualize future cabin installations in AR mode. When showcased, the AR technology was used to visualize location and shape of new components in the aircraft cabin. The ability for this project to quickly switch demonstrations between commercial and business cabins in various configurations was admired by many users.

5 Future Outlook and Road Map

5.1 Pilot Training Module Using Augmented Reality

The MIMS team is currently working on redeveloping the pilot training tutorial to improve performance and Field of View (FOV) limitations associated with the HoloLens. The HoloLens is limited to a 35–40° FOV, meaning any information placed or located outside of that range would not be viewable by the pilot in the neutral head position. The objective of the tutorial is to provide supplementary information without increasing the workload of the pilot. As a result, the pilot should not need to move their head around in space to look for tutorial information; rather they should be focused on safely completing the take-off manoeuvre. For this reason, the improved tutorial will place all graphical instructions within the FOV of the HoloLens (Chittaluri et al., 2017).

5.2 Interactive Virtual Aircraft Cockpit

The MIMS team considers this project to be complete as it achieved its objectives. No additional work on this project is planned at this time. However, the team is using this project to promote VR technology use in rapid prototyping, simulation, and training. The project is also used as an example of VR capabilities in improving cockpit development, pilot training, and other aerospace operations.

5.3 Head-Mounted Display Using 2D and 3D Views for Aircraft Flight Simulation

Future outlooks for this project include using the developed BLT communication protocol to develop and test cockpit HMI^2. Concern for clutter and symbology representation raised by some users may be addressed in future iterations. Cognitive ergonomics and HMI^2 design theories will play an important role in optimizing the HMD display information and symbology. The second-generation Microsoft HoloLens will be used for future iterations of this project.

5.4 Augmented Reality Technology for Visualization of Aircraft Interior Components

The visualization segment of this project is complete. The MIMS team is focusing cabin research efforts on developing the components and layouts visualized with AR in this project. The team is investigating the use of transparent screens on board aircraft to create smart cabin class dividers and interactive window displays. These concepts are being developed with a host of industry partners and organizations with the objective of improving the aircraft cabin. Future phases of the cabin interiors research at the MIMS lab are largely concerned with layout and HMI^2 optimization.

6 Conclusion

Ongoing research at the MIMS lab uses MR devices to enhance the flight experience for passengers and crew members. The various projects initiated at the lab have proven the viability of MR use in the aerospace sector for purposes beyond flight data augmentation. MR devices can be used for crew training, rapid development of aircraft cockpits, and visualization of cabin interiors. Users and stakeholders are supportive of the work done at the MIMS lab. The large amounts of positive feedback received for each project discussed in this chapter is a good indicator of the aerospace industry's willingness to use MR in daily operations. The current abilities of COTS hardware and software imposes some limitations on the extent of MR use in the field, though it is expected that this technology will improve tremendously over the next few years. These developments are necessary for future success of projects at the MIMS lab, where the research team is working to enhance, test, and apply existing projects in various aerospace operations and tasks.

References

Austakalnis, S. (2017). *Practical Augmented Reality: A guide to the technologies, applications, and human factors for AR and VR*. Pearson Education.

Billinghurst, M., Clark, A., & Lee, G. (2015). A survey of augmented reality. *Foundations and Trends in Human–Computer Interaction, 8*(2–3), 73–272.

Chittaluri, V. (2017). *Development of first person view for a UAV ground control station with the Microsoft hololens*. Ryerson University.

Chittaluri, V., Chung, J., Pradhan, P., & Park, H. U. (2017). Development of holographic user interface for UAV ground control using Microsoft HoloLens. *Canadian Aeronautics and Space Institute Aero 2017 Conference.*

Frigo, M. A., Silva, E. C., & Barbosa, G. F. (2016). Augmented reality in aerospace manufacturing. *Journal of Industrial and Intelligent Information, 4*(2), 125–130.

MIMS. (2016). *Welcome to the MIMS Lab at Ryerson*. Retrieved February 10, 2020, from https://people.ryerson.ca/j3chung/index.html

Morrison, M. (2018). *Inside stories from the Crystal Cabin finalists*. Flight AIX Daily News, 10 April, p. 34.

Peddie, J. (2017). *Augmented reality: Where we will all live*. Springer.

Safi, M., Chung, J., & Pradhan, P. (2019). Review of augmented reality for aerospace industry. *Journal of Aircraft Engineering and Aerospace Technology, 91*(9), 1187–1194.

AR and VR to Enrich Cultural Heritage and Retail Experiences: ETT Case Studies and Guiding Principles

Adele Magnelli, Davide Pantile, Matteo Ventrella, and Giovanni Verreschi

Abstract

In recent years, both the cultural and the industrial/retail sectors have been going through an intense period of development, characterized by suggestions from ICT. The use of Augmented and Virtual Reality features for the enjoyment of cultural and retail experiences has become increasingly demanded when it comes to engage different users and increase levels of immersion aiming to the transmission of information. ETT, specialized in technological innovation and Experience Design, has developed high level know-how in the creation of AR and VR experiences, both in Culture and Retail, though the use of cutting edge technologies and engagement and storytelling techniques.

Keywords

Augmented reality · Audience engagement · Immersive experience · Virtual reality

1 Company Description

ETT S.p.A. is a Digital and Creative Industry specialized in technological innovation and Experience Design. It was established in 2000 and it currently employs over 150 people between Genoa headquarters and various offices in main Italian cities and in London. ETT has been active for years in the New Media sector, for which it creates innovative applications exploiting the potential of new technologies in contexts related to edutainment, culture, tourism, communication and marketing. It

A. Magnelli (✉) · D. Pantile · M. Ventrella · G. Verreschi
ETT S.p.A, Genoa, Italy
e-mail: adele.magnelli@ettsolutions.com; davide.pantile@ettsolutions.com; matteo.ventrella@ettsolutions.com; giovanni.verreschi@ettsolutions.com

T. Jung, J. Dalton (eds.), *XR Case Studies*, Management for Professionals,
https://doi.org/10.1007/978-3-030-72781-9_7

combines original design, storytelling and cutting-edge technologies to create engaging experiences for museums, corporate and retail spaces. In this fields, since 2008, ETT has delivered over 1000 multimedia installations in about 100 museums and private customers, for a total of over 4.5 million visitors. Even if ETT case histories concern application areas that at first glance may seem distant—Cultural Heritage and Industry/Retail—the approach that guides project creation and experience development is unitary. ETT's aim is to create unique user experiences, focused on engagement, in order to improve the level of "knowledge"—whether it is about the cultural attractor or the product/industrial process—by using AR and VR immersion techniques, interactivity and storytelling.

2 Project Summary

2.1 Augmented Reality Projects

ETT has designed and developed many Augmented Reality applications, allowing users to explore cultural heritage assets and product features in an interactive way. In Augmented Reality applications, using the video stream captured by the mobile device camera and associated to image recognition, it is possible to add multimedia contents and information (e.g. texts, pictures and 3D reconstructions) on the surrounding real world or object. AR applications developed by ETT allow users to explore virtual environments, view the 3D reconstructions of ancient monuments overlaid on their real ruins, interact with 3D Avatars of historical figures. In the following paragraphs the main projects presented are: Cenacolo Vinciano App, Accademia Galleries App and UnipolSai CarViewer App.

2.2 Virtual Reality Projects

Focusing on the enhancement of cultural heritage, smart tourism, corporate spaces and retail, ETT has designed and developed many immersive tools based on Virtual Reality features. Virtual Reality is used to replace the user's real-world environment with a simulated one. Using specific VR headsets, users can enjoy immersive virtual tours of fascinating historical sites or reproductions of industrial processes, with means of both entertainment and training. Thanks to Virtual Reality applications, users can explore virtual environments by feeling completely immersed in 360-degrees panoramic views or computer graphics 3D reconstructions of sites and scenes. In the following paragraphs the main projects presented are: Ars Excavandi, Teatro alla Scala Museum, Area X—Intesa San Paolo Assicura.

2.3 Projects Offering Both Augmented and Virtual Reality Experiences

Furthermore, many ETT projects encompass both Augmented and Virtual Reality features, in order to create immersive experiences with many degrees of immersion and engagement. In the following paragraphs the main projects presented are: The Ara as It Was, Virtual Destination Italy, Exploracity and Musei in Comune—Interactive Video Guide.

3 Project Details

3.1 Augmented Reality Projects

In the Cultural Heritage field, the main AR including projects are:

Cenacolo Vinciano App
The new Leonardo da Vinci's Last Supper App makes the visit experience to the Museum even more comprehensive and engaging. The free App, developed both for Android and iOS devices, takes museum visitors on a unique tour. The Last Supper's charm is revealed through narration techniques, helping visitors to appreciate the technical and compositional choices made by Leonardo. Inside the Refectory, the AR feature makes the visit even more immersive and evocative. The Stories section is dedicated to further historical and artistic information that helps to contextualize the painting through text cards supported by 360-degree videos and pictures.

Accademia Galleries APP
In the monumental estate of the Academia Galleries of Venice, ETT redesigned the visit experience by several touch exhibits and a mobile application, in order to improve the visitor experience and enjoyment of works of art. The mobile app, developed for iOS and Android platforms, is designed for different types of users—Children, Teenagers, Adults—and helps investigate aspects and details of the works that cannot be perceived at first sight. Augmented Reality contents, through which visitors can see what emerged during restoration work and the artworks in three-dimensional mode, in addition to videos with in-depth information, are available.

In the Retail, the most recent AR including project is:

UnipolSai CarViewer App
CAR Viewer is an Augmented Reality application that offers an innovative solution for discovering vehicles, exploring them virtually, deepening knowledge of the various models characteristics through 3D Real Time vehicle reconstructions. Clients can assess different vehicles until they find the ideal one, before they even touch it, in a simple, immediate and safe way. The application allows to visualize and interact with the model and to position it in the room where the client is, maintaining

contact with the real world. The AR experience can be enjoyed through last-generation tablets or smartphones.

3.2 Virtual Reality Projects

In the Cultural Heritage field, the main VR including projects are:

Ars Excavandi

"Ars Excavandi" is the exhibition that opened the year of Matera 2019 as the European Capital of Culture. Thanks to visitors take part to an immersive excursus into the history of underground art and relive the magnificence of the excavated architecture. The final part of the exhibition is situated in the hypogea, characteristic underground places of the city of Matera. They're 1200 sq. Meters underground that can go up to a depth of 12 meters. The Virtual Reality experience developed by ETT mimics a walk in the hypogea and guarantees the chance to visit the "underground Matera" even to those who are affected by physical disabilities and could not otherwise enjoy it.

Teatro Alla Scala Museum

The digital innovation project developed by ETT is related to the strategy of the Teatro alla Scala Museum for the enhancement of the exhibited works. Besides multimedia stations and a mobile app guiding visitors along the Museum itinerary, the novelties include, in particular, a Virtual reality tour of the Theatre. Wearing specific VR headsets and guided by the prima ballerina Nicoletta Manni, visitors can explore the Opera Theatre and discover its beauties and secrets in a highly engaging tour.

In the Retail field, the most recent VR including project is:

Area X: Intesa Sanpaolo Assicura

Area X is an innovative edutainment space that relies on the concept of "protection", where live interactive and engaging experiences in Virtual Reality take place. Crossing the doors of the Intesa San Paolo exhibition area, a unique adventure starts, bringing the visitor to an evocative alien planet where protection is as important as on planet Earth. Here, as settlers, visitors are asked to choose how to act in this new world. In three different stations is possible to explore the planet in different environments: driving a jetpack to fly among suggestive floating islands, visiting an alien housing module or driving a space rover on a journey through breath-taking 3D scenarios. The experiences are linked by one guiding principle: the more you are protected, the more fascinating and full of surprises the adventures will be. Before facing an adventure, special "space" insurance policies may be selected and taken on the trip. These will protect the score from the unforeseen events that would lower it. This helps people to understand which is the best type of insurance for every circumstance in life.

Fig. 1 The Ara as It Was, *Rome*

3.3 Projects Offering Both Augmented and Virtual Reality Experiences

The following projects include both Augmented and Virtual Reality features, and have all been developed in Cultural Heritage and Tourism fields.

The Ara As It Was

The "Ara as It Was" experience is the first systematic Augmented and Virtual-reality project carried out in Rome about Cultural Heritage. The Ara Pacis is one of the most important Roman art masterpieces, built to celebrate the peace achieved by the Emperor Augustus throughout the Roman Empire. Using Samsung Gear VR headsets and Samsung S7 smartphones, a multimedia story unfolds, showing the original appearance and function of the altar, both using AR and VR features. In the first part of the visit, it is possible to see the alterations that happened through time to the northern part of Campo Marzio, the area of Rome chosen by Augustus to acclaim his power, through Virtual Reality. The second part of the experience is instead based on Augmented Reality. Using the device camera, it is possible to activate contents about the original colours of the monument and the stories represented by the bas-relives. Thanks to an algorithm developer by ETT, the multimedia contents are anchored directly on the monument, without ever interrupting the vision of the real monument (Fig. 1).

Virtual Destination Italy

Virtual Destination Italy is an innovative platform offering a multisensory interactive tour in which the visitor can identify himself with the character of a young nobleman of the XVII century, taking his Grand Tour in Europe and Italy. The Grand Tour is the traditional trip of Europe undertaken by mainly upper-class European young men. The Virtual Destination Italy platform, thanks to a mobile application and geolocation tools, offers immersive experiences, where the user interacts with various characters, scholars and artists of the time that, through Augmented Reality features, appear on the screen in the form of 3D avatars and tell stories and anecdotes about the different places. The platform, moreover, offers visitors Virtual Reality experiences, dynamic comparisons between past and present pictures of the same place and 360-degree panoramic images, in order to enhance the knowledge of the Cultural Heritage offer of the different places that once made the Grand Tour great.

Exploracity

Exploracity is an innovative platform for the enhancement of tourism and cultural heritage of the city of Genoa. As a pre experience, Exploracity comprises touch screen stations and VR experiences, located at the Tourist Information Offices, offering multimedia contents, photos and videos about the main points of interest of Genoa, with gaming applications dedicated both to children and adults. During the experience itself, an interactive and immersive exploration of the city is made available to users thanks to a mobile Application that, based on GPS geolocation tools, offers amazing Virtual and Augmented Reality experiences. Thanks to this platform, visitor receive the first notions and historical reconstruction of the city's cultural development.

Musei in Comune Rome: Interactive Video Guide

The Civic Museums of the City of Rome, Italy—Musei Capitolini, Museo dell'Ara Pacis and Mercati di Traiano-Museo dei Fori Imperiali—have been equipped with a highly innovative video guide created by ETT, which combines the characteristics of a traditional audio guide with innovative technologies, thereby creating a "multimedia tour". The interactive video guides aims to revive the visiting experience through the curiosity and emotional engagement, achieved with an indoor location system—based on Beacon (BLE) technology—that automatically suggests content to visitors that is associated with nearby artwork. Augmented and Virtual Reality also support the delivery of information content via storytelling techniques, by showing, for example, the protagonists of the frescos (Fig. 2), X-rays of paintings and 360° photos of the museums halls.

4 Feedback from End Users

Even if the main user feedbacks that ETT projects receive are positive—in terms of usability, engagement, transmission of information—"The Ara as It Was" at the Ara Pacis Museum in Rome is the main project developed by ETT that has been analysed and tracked by an academic committee, in partnership with Università degli Studi di Napoli Parthenope (Italy). The latest empirical analysis has been carried out applying the visitor experience model for mixed reality, which measures how 23 functional

Fig. 2 Musei in Comune Interactive Video Guides, *Rome*

and experiential elements have been reshaping a traditional museum visit, enhancing visitor satisfaction (Trunfio et al., 2019). The empirical analysis findings and conclusions open up new scenarios for future research on immersive museum experiences, combining cultural heritage with technological innovation. One of the latest paper produced on the project "The Ara as It Was" tries to shed light on this topic by analysing how mixed reality functional and experiential elements have been reshaping the traditional museum visit, opening up opportunities and challenges to build innovative immersive experiences and enhance visitor satisfaction. The Ara Pacis Museum constitutes a best practice of heritage exhibition management under conditions of mixed reality, in which virtual interaction with cultural and historical artefacts enhance the immersive visitor experience (Bec et al., 2019). By integrating physical museum elements with virtual and augmented spaces, the "Ara as It Was" project reshapes the iconic Ara Pacis Museum and redesigns museum service models. Empirical analysis results show that the "Ara as It Was" project represents a successful combination of advanced technology and innovative storytelling, increasing visitor presence and participation levels. Furthermore, findings of the empirical analysis allow the introduction of fresh knowledge into the theoretical debate and propose some managerial implications for the co-creation of value in the museum experience.

5 Future Outlook/Roadmap

As the media theorist McLuhan pointed out in 1960, media types are way more than just passive channels of information, being capable of affecting people in different ways. Even if it is still not completely known what extent AR and VR will impact human thinking processes, it is clear that the use of these technologies has become an integral part of cultural, retail and training experiences. The main perspective that is expected in the field of AR and VR technologies is certainly the implementation with Artificial Intelligence mechanisms, which can be used, in the form of chatbots or natural language comprehension, to enhance and enrich user experience.

6 Conclusion

Augmented and Virtual Reality features development is changing Cultural Heritage and Retail realities impressively fast. This process allows Industries working in ICT, like ETT, to transform the relationship between users and experiences. Users turn the visit into an increasingly personal journey of contamination, shared emotion and involvement. The constant evolution of the needs of the public—both Cultural Heritage and Retail—is based on the search for experiences and new stimuli. An answer to the search for these new stimuli can be represented by the wise use of AR and VR technologies, which ETT has consolidated thanks to the know-how developed in the field. The wise use of AR and VR technologies is related to the transmission of correct and scientifically validated information, and to the use only in cases where they enrich the experience in ways that would otherwise not be possible. In conclusion, it is clear that the use of technologies is destined to improve—in terms of usability and success in conveying information—if guided by constant research and comparison with international scenarios development. These are ETT's guiding principles.

References

Bec, A., Moyle, B., Timms, K., Schaffer, V., Skavronskaya, L., & Little, C. (2019). Management of immersive heritage tourism experiences: A conceptual model. *Tourism Management, 72*, 117–120.

Trunfio, M., Campana, S., & Magnelli, A. (2019). Measuring the impact of functional and experiential mixed reality elements on a museum visit. *Current Issues in Tourism*. https://doi.org/10.1080/13683500.2019.1703914

Augmented Reality Using ViXAssist and HoloLens 2 for Automotive Service and Maintenance

Henry Chow

Abstract

The cost and complexity of service and maintenance are increasing as vehicles designs become more sophisticated. Holovis developed an AR maintenance system for a luxury vehicle OEM using the HoloLens2 and VixAssist to support service engineers at dealerships worldwide. The system is hands-free, allowing engineers to carry out service and maintenance tasks. Annotations can be added to selected Points of Interests (POI) via a virtual keyboard. Annotations are automatically uploaded to a remote server, which can be accessed by a separate third-party PLM or project management system to facilitate data integration. Users can select a POI and access relevant technical information, which is displayed as a floating window within the environment. This simplifies information access and eliminates the need for printed manuals. Holovis demonstrated that the technology has reached a level of maturity for early adopters. However, there remains a number of technical and ergonomics challenges that need to be resolve for widespread field operations.

Keywords

Augmented reality · Automotive · Wiring harness · Gesture recognition · AR · HoloLens

H. Chow (✉)
Holovis International Limited, Hinckley, UK
e-mail: henry.chow@holovis.com

T. Jung, J. Dalton (eds.), *XR Case Studies*, Management for Professionals,
https://doi.org/10.1007/978-3-030-72781-9_8

1 Company Description

Holovis International Ltd. is a world leading innovator and solution provider of sensory and extended experiences by bringing together leading-edge emerging enabling technologies. Holovis HQ is located the Midlands UK with offices in London, Manchester, Orlando and Shanghai.

With over a hundred technologists worldwide across multiple disciplines, Holovis delivers end to end experiential solutions across industries. Within entertainment, Holovis is recognised as a global leader in themed entertainment with clients such as Merlin Entertainment, Sea World and Farah Leisure. By combining creative story telling with immersive multi-sensory technologies, Holovis creates some of the most compelling experiences, transforming some of the world's most popular visitor attractions.

For enterprise applications, Holovis has a track record of innovation within automotive and manufacturing with a portfolio of high-profile clients, such as McLaren, Jaguar Land Rover and BAE Systems. With industrial subject matter expertise, Holovis integrates cutting-edge technologies such as VR/AR, tracking systems, computer vision and big data systems. Combined with creative and engaging user experience designs, Holovis expands the way human interacts with industrial applications from industrial/product design and training, to quality monitoring and machine maintenance.

Holovis continues to innovate. The company is engaging in a R&D initiative that combines real time data acquisition and machine learning with Virtual and Augmented Reality use interfaces to rethink the way we interact with complex data and information for both entertainment and enterprise applications.

2 Project Summary

Vehicle design and automotive electronics are becoming increasingly complex. The 1927 Ford Model T has 1481 parts, while a modern vehicle has over 30,000 parts (Japan Management Association, 1986) with some 1500 copper wires totally about 1 mile in length (Sprovieri, 2014). This is inevitable as technology progresses alongside increasing competition, regulations and customer expectations. A consequence of this complexity is the increase of service and repair costs (Gotsch, 2018). The automobile has become part of a larger service system, encompassing *"dealers, authorised repair shops, rental organisations and the users themselves"* (Singh, 2010).

To help mitigate escalating service and maintenance costs, a luxury vehicle OEM (Original Equipment Manufacturer) in the UK wanted to explore the adoption of Augmented Reality for the service and maintenance of its vehicles. Holovis embarked on a Proof of Concept (PoC) that seeks to tackle three technical challenges:

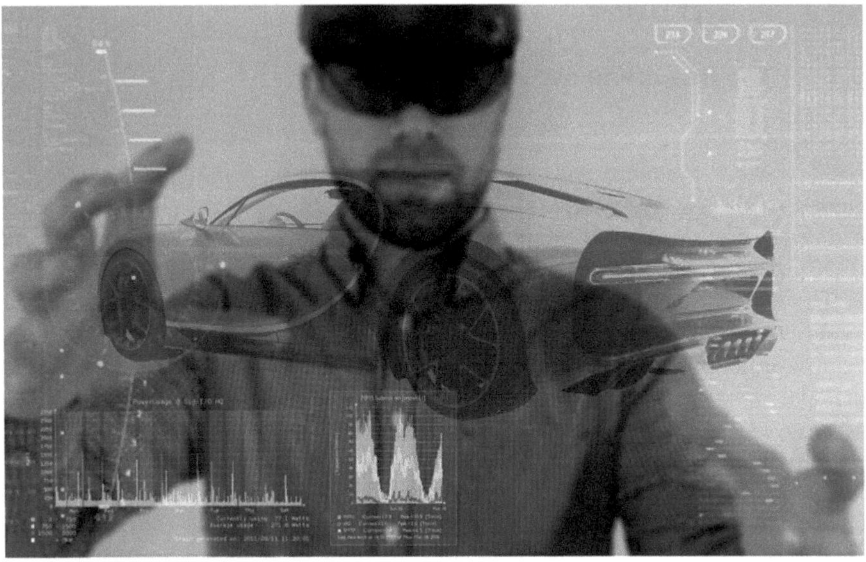

Fig. 1 Augmented Reality HoloLens 2

- 2D, digital or printed, wiring diagrams do not adequately communicate the three-dimensional geometry and routes of a vehicle wiring harness in contextual relationship to the vehicle form;
- access to technical documentation and information can be difficult during inspection and maintenance tasks; and
- maintenance tasks and notes are manually written afterwards or on separate data systems requiring duplicate data entry, which is prone to error and omission.

Holovis developed an Augmented Reality maintenance system called ViXAssist. Using the Microsoft HoloLens2, the wiring harness of vehicle is superimposed onto the physical vehicle through the HoloLens' holographic lenses, known as waveguides. This enables the engineer to see exactly where the wires and connectors are intuitively. In so doing, the engineer can access, disassemble and reassemble the vehicle without the need for external printed references which can be cumbersome.

Via the HoloLens2 hand gesture interface, service engineers can select parts of the wiring harness to access detailed technical information. Moreover, service engineers can add annotations that are associated with a specific component. Annotations can be general comments, notes or outstanding tasks. Meta data, such as date, time and engineer ID are automatically associated with each annotation, which are logged in real time onto a central data server. Data can then be disseminated into a project management and PLM systems, and can be reviewed at a later date Figure 1 illustrates how a user can interact with the AR environment, using the HoloLens 2 with data overlay and gesture recognition.

3 Project Detail

3.1 The Requirements

As vehicle design are becoming increasingly complex, a luxury brand automotive OEM wanted to explore the use of emerging enabling technologies to support engineers at their dealerships globally and help mitigate escalating service and maintenance costs. The OEM approached Holovis to conduct a Proof of Concept (POC) to assess the viability of different technologies.

The fundamental premise of the project is to incorporate the use of Augmented Reality (AR) to correctly and accurately align the digital twin wiring harness of the vehicle on its physical counterpart. The system must also be able to visualise the digital twin independent of the physical model.

From a UI (User Interface) perspective, the system must be hands free with audio cues to enhance interactions. The system must incorporate hotspots or Points of Interests (POI), with which the user can interact. The system must be connected to a Content Management System (CMS) for the user to retrieve up-to-date information pertaining to the POIs. Content includes documentation, in the form of PDF or HTML, and videos. To help users navigate around the complexity of the vehicle, the user needs to have the ability to toggle the visibility of components, for example, to only view the electrical components. In addition, the system must incorporate the ability to view real time 3D animations, for example, the assembly/disassembly of components.

A key requirement is to allow the user to make annotations that are attached to specific POI of the vehicle. Annotations can be both text and voice notes and are uploaded to a remote server to support the wider PLM process.

Since the technology seeks to target global dealerships, there are two additional functional requirements. Firstly, the system should be multilingual. Secondly, the system should incorporate a remote assistance feature that allows engineers at dealerships worldwide to contact specialists in the UK.

3.2 The Technology

Holovis integrates four pieces of technology for this project.

VixAssists is Holovis' proprietary Augmented Reality (AR) system designed for manufacturing and maintenance use cases. It originated from iGuide, which is an AR vehicle owner's manual. The ambition of iGuide was to eventually replace the printed owner's manual. Using the AR interface on a smart phone, iGuide provides operating instructions around the vehicle centre console and dashboard. iGuide was generalised into ViXAssists, which is device agnostics to work on smart phones, tablets as well as laptop/desktop computers.

ViXCore is a sub-component of DeepSmarts, which is Holovis data *compute, analytics* and *visualisation* platform. ViXCore facilitates real time data communication between the AR device, the CMS and a remote database, which in turn is accessible from a PLM system.

The Microsoft HoloLens2 was selected over other AR devices to provide a complete hands-free experience. It has built in hand tracking, eye tracking as well as voice control (Microsoft, n.d.). According to Alex Kipman, the HoloLens 2 waveguides delivers a 52° diagonal FOV (Wired, 2019), which gives an approximate horizontal FOV of 43° and a vertical FOV of 29°. This is 43% and 65% larger than the original HoloLens horizontal and vertical FOV respectively (Heaney, 2019).

The AR environment was developed using Unity as the core rendering engine and Microsoft's Mixed Reality ToolKit V2 was used for the AR component of the system.

3.3 The Challenges

While the project in its entirety was uneventful, we did encounter some challenges on the way. When the project commenced, Holovis was fortunate enough to have early access to the HoloLen2. This did mean that documentations, SDKs and community support were not widely available and sparse, compared to existing devices which ViXAssists was built upon. Since the MRTKv2 was new, we encountered a number of bugs, which meant that a lot of testing had to be done on the device itself.

A lot of changes were made throughout the development process to accommodate a wider range of users. Since ergonomics is one of the crucial factors of the development, a number of customisation functions were added. Although the majority of them were simple, they are not usually required in typical POC projects. A tutorial, similar to what you might find in a computer game, was introduced, allowing users to "learn and try" to achieve device familiarity and to help users get used to a new way of working.

4 Feedback from End Users

Overall, the technology integrates and operates as anticipated. HoloLens 2 worked well and was remarkably comfortable to use compared to the original HoloLens. It has a flip up visor and can accommodate users with spectacles. The improved FOV was significant to provide a much more conducive environment for service and maintenance applications.

Hand tracking and gesture recognition allows users to select components and retrieve documentation without the need for a handheld device. Using the built-in virtual keyboard, users can enter text notes and tag them to specific components, again without the need for a wand or other handheld devices, leaving their hands free to carry out physical tasks.

The POC demonstrated successfully to the OEM leadership team that the technology is viable and has reached a level of maturity for field implementation into global dealerships as a pilot for early adopters. However, a number of ergonomics issues remain to be resolved.

The virtual keyboard is still cumbersome to use, making data entry a slow process when compared to a conventional set of keyboard and mouse. As speech to text becomes increasingly accurate and common place, it is anticipated that in the near future, the keyboard may become obsolete (Brown, 2019).

While gesture recognition worked well, users need to be very deliberate in order to invoke the correct command. Since the HoloLens is a new piece of technology, the interface is still not widely intuitive for many new users. Therefore, it is imperative that the system incorporates a robust and engaging training feature to reduce its learning curve.

Users can quickly access technical documents in the form of a PDF or HTML, which are displayed as a floating window within the AR environment. However, they can be difficult to read within the AR environment. Users can move, resize and even walk up closer to it. However, this is not intuitive compared to just picking up a printed document and reading it.

An interesting observation is that, while users can retrieve technical documents by selecting a component, in reality, engineers often need to do the reverse where they need to identify the physical location of a component from the technical document. Thus, a bi-directional link is advantages between technical documentation and the AR environment.

5 Future Outlook/Road Map

Augmented Reality is undoubted a very promising technology. With the commoditization of smart phones and the continual development and miniaturisation of wearable AR devices, AR's growth is expected to surpass that of VR (Statista, 2020). "VR will be big, AR will be bigger" (Merel, 2017). The arrival of 5G may help pave the way for miniaturisation by off-loading computation and rendering processes to the cloud (Marr, 2020).

Two obstacles remain.

Firstly, a main obstacle is the usability of deviceless interaction necessary for deskless application, such as VR and AR. While gesture and speech can be intuitive at first, accuracy and precision are difficult to achieve compared to the keyboard and mouse. Baig and Kavakli (2018) found that "*efforts for performance*" and "*fatigue levels*" for gesture and speech are much higher than those of the keyboard and mouse. "*The usability of traditional input devices* [keyboard and mouse] *supersedes the multimodal inputs* [speech and gesture]." However, Baig and Kavakli postulate that "*the gap between these two has been minimized, and with advanced technology for speech and gesture recognition, it can be overcome in near future.*" With sophisticated AI and/or fuzzy logics, it is possible for computer systems to make inferences, allowing users to issue complex and precise commands without or with less necessities for precision input devices, like the keyboard and mouse.

The second is the cost of content creation. Currently, the creation of digital 3D content with the necessary behaviours and triggers is time consuming and labour intensive. Content creation typically requires a team of SW developers and artists, creating an expensive bottleneck (Porter & Heppelmann, 2017). Outside of high value manufacturing, this can be cost prohibitive.

Since Holovis is not an AR hardware manufacturer, we are reliant on the likes of Microsoft and Magic Leap. But the technology is maturing quickly and is already robust enough for early adopter applications.

To help mitigate the cost of content creation, Holovis is developing a Content Generator that will allow the OEM to add POI and link technical documents to the AR model. This means that once the AR environment has been set up, the OEM has the ability to edit the models. In addition, Holovis will implement a vehicle configurator by integrating our Rule Engine Services (RES), allowing the OEM to create different variants of a vehicle platform. An automated process will allow a service and maintenance engineer to identify a specific car through its unique VIN and visualise its exact specification, configuration as well as service history.

Holovis also plans to implement a remote collaboration tool through integration with a Video Streaming Services (VSS). This allows engineers abroad to contact specialists in the UK HQ during a maintenance task. Unlike standard video conferencing, the remote user will be able to see exactly what the engineering is seeing and interact with the digital models.

Holovis is also exploring the implementation of Bluetooth, allowing the HoloLens 2 to connect to an iPad. This has three potential benefits. Firstly, technical documents can be viewed on an iPad, which is much easier to read. Secondly, the iPad can be used to add annotation in lieu of the AR virtual keyboard, which will make typing simpler. The iPad may also be able to serve as a control interface. While this contravenes the principle of deviceless interaction, this may be a suitable way to mitigate the low level of usability with today's hand tracking and gesture recognition technology.

6 Conclusion

The POC demonstrated that AR technology has reached a level maturity ready for field implementation. While challenges remain, in particular around the cost of content creation and the ergonomics of deviceless interaction, Augmented Reality has shown to be a promising technology to support engineers and to help mitigate the escalating cost of vehicle service and maintenance. Data connectivity allows users to access data and information through the deviceless AR interfaces and supports a much more integrated PLM process.

Moving forward, a Content Generator will help reduce the cost of content creation and custom configuration. A real time video streaming feature will enable engineers around the world to collaborate effectively by allowing them to see what each other sees across different devices. iPad or tablet Bluetooth integration can serve as a stop gap to mitigate existing usability issues.

References

Baig, M. Z., & Kavakli, M. (2018). *Qualitative analysis of a multimodal interface system using speech/gesture*. Irtual and Interactive Simulations of Reality (VISOR) Research GroupDepartment of Computing, Faculty of Science and Engineering Macquarie University, Sydney, Australia.

Brown, E. (2019). *Will keyboards be obsolete soon, as voice tech takes over?* Retrieved from ZDNet: https://www.zdnet.com/article/will-keyboards-be-obsolete-soon-as-voice-tech-takes-over/

Gotsch, S. (2018). *Vehicle complexity driving increases in repair time cycle*. Retrieved from https://www.claimsjournal.com/news/national/2018/07/16/285736.htm

Heaney, D. (2019). *HoloLens 2's field of view revealed*. Retrieved from Upload: https://www.wired.com/story/microsoft-hololens-2-headset/

Japan Management Association. (1986). *Kan Just-In-Time at Toyota*. Productivity Press.

Marr, B. (2020). *The 5 biggest virtual and augmented reality trends in 2020 everyone should know about*. Retrieved from Forbes: https://www.forbes.com/sites/bernardmarr/2020/01/24/the-5-biggest-virtual-and-augmented-reality-trends-in-2020-everyone-should-know-about/#228521af24a8

Merel, T. (2017). *The reality of VR/AR growth*. Retrieved from https://techcrunch.com/2017/01/11/the-reality-of-vrar-growth/

Microsoft. (n.d.). *HoloLens 2*. Retrieved from Microsoft: https://www.microsoft.com/en-us/hololens/hardware

Porter, M. E., & Heppelmann, J. E. (2017, November–December). Why every organization needs an augmented reality strategy. *Harvard Business Review*.

Singh, N. (2010, February). *Cars and complexity*. Retrieved from https://www.livemint.com/Opinion/GEwzd0rCq9qPFuKMOIv5YP/Cars-and-complexity.html

Sprovieri, J. (2014, July). *Wire harness recycling*. Retrieved from https://www.assemblymag.com/articles/92263-wire-harness-recycling

Statista. (2020). *Augmented reality dossier*. Retrieved from www.statista.com

Wired. (2019). *Microsoft's HoloLens 2 puts a full-fledged computer on your face*. Retrieved from https://www.wired.com/story/microsoft-hololens-2-headset/

Viarama: Virtual Reality for Good

Billy Agnew

Abstract

Viarama is a social enterprise that uses virtual reality to benefit schoolchildren and senior citizens in schools, nursing homes, hospices, hospitals, and respite centres, mainly in the UK. This paper explores the issues faced by this work and the often deeply moving nature of utilisation of VR in these environments.

Keywords

Hospices · Nursing homes · Schools · Respite centres · Hospitals · Senior citizens · Schoolchildren · Healthcare · Education · VR for good

1 Company Description

1.1 Viarama C.I.C

Established in 2015, Viarama is a world-leading social enterprise, based in Edinburgh, that has been delivering VR sessions in schools, nursing homes, hospices, respite centres, and hospitals in Scotland since 2016. In addition to this core work the company creates VR experiences for various organisations, all of which have a strong social focus. Viarama uses room-scale virtual reality, with HTC Vive headsets at all of our sessions.

Their work is multi-faceted but the main focus is upon using VR to improve quality of life among schoolchildren and senior citizens. Engaging with children and adults with all levels of physical and cognitive ability, their work involves people living with a wide variety of physical and mental conditions, including Alzheimer's

B. Agnew (✉)
Viarama C.I.C., Edinburg, UK
e-mail: billyagnew@viarama.co.uk

T. Jung, J. Dalton (eds.), *XR Case Studies*, Management for Professionals,
https://doi.org/10.1007/978-3-030-72781-9_9

disease, dementia, Asperger's syndrome, autism, multiple sclerosis, and many other conditions.

Viarama utilises a diverse range of VR applications and experiences, as required, and has developed a range of methodologies from both a psychological and technical perspective to allow clients to enjoy the experiences offered to their maximum potential.

All work utilises a person-centred approach to engage with clients on a one-to-one basis, and compassion is at the core of its ethos, and this approach underpins efforts in all areas.

2 Project Summary

This case study will look at Viarama's work delivering VR sessions in schools, nursing homes, hospices, respite centres, and hospitals.

All of these areas of operation differ, often greatly, and the approach taken has been tailored to each setting based upon extensive experience gained since 2016.

3 Project Details

3.1 VR in Education

Viarama works with pupils in both mainstream and special needs education, and includes work with pupils receiving little to no formal education due mainly to behavioural problems. This study will firstly focus upon this group.

Pupils in the group present with one or more of the following issues:

- learning difficulties;
- significant behavioural problems;
- social difficulties;
- experience of one or more adverse childhood experiences;
- mental health issues;
- developmental disorders.

In this use case, VR is used as a tool to improve:

- self-esteem;
- general confidence;
- problem-solving skills;
- social skills;
- self-expression;
- resilience;
- general wellbeing and overall mood;
- desire, willingness, and ability to learn;

- willingness to try new things;
- general performance in school.

Viarama does this by establishing a rapport with each pupil, gaining their trust where possible, and ascertaining their interests, desires, likes and dislikes based upon the information they volunteer, and that of their teachers and carers. From there they begin to tentatively introduce each pupil, on a strictly one-to-one basis, to VR; being very careful to ensure the initial experience of the technology is relaxed, friendly, non-threatening, and completely free of pressure to either try VR in the first instance, or continue with it. This gentle approach has been very successful with almost all of the several thousand pupils we have worked with to date. An overwhelming majority of pupils reported a positive response to the sessions, with documented feedback received as follows:

- would very much like to try the experience again and wanted to know when or if this could happen;
- was delighted, enthused, and very excited to have tried VR;
- was thrilled to be able to revisit their previous home following immigration to Scotland;
- was stimulated to learn more about the topic or topics covered;
- learned an interest they didn't know they had;
- reignited a spark of enthusiasm for something that had fallen by the wayside;
- were soothed or calmed by a relaxing environment;
- expressed enthusiasm to stay in VR;
- expressed enthusiasm to learn to develop, code, or produce VR content;
- expressed a desire to work in a career path for the first time;
- experienced joy, wonder, excitement, positivity, and happiness at the experience.

The above is by no means an exhaustive list of the positive emotions experienced or expressed by pupils Viarama has worked with. Unprompted, and in the vast majority of instances, pupils expressed a strong desire to do more, learn more, and enjoy more VR sessions. This enthusiasm was conveyed to their teachers on a great many occasions, and was used as a starting point for creative lessons of different kinds. It was noted that for several pupils this was the first time they had expressed enthusiasm for this type of schoolwork.

It was decided that Viarama should attempt to measure the responses and track the overall mood among pupils prior to and after their VR session. To that end a simple form was created that would capture qualitative and quantitative data that could subsequently be used to measure pupil responses in detail.

Pupils were encouraged to speak freely and with candour about their experience of VR. The questions were deliberately left open to interpretation; the reason being that the pupil would unconsciously decide the focus of the questions and respond accordingly.

From these completed forms, a hugely positive picture of the use of VR in education has formed. Of all participants, there was a 100% positive response from how the pupils felt upon arrival to how they felt at the end of the session. *All pupils* felt as good or better after they had enjoyed VR.

Pupils who had previously expressed suicidal ideation no longer did so. Pupils who exhibited very low self-worth or self-esteem were beginning to believe in themselves to a greater extent. Resilience and tenacity in the face of problems showed a marked improvement across the entire group. Problem-solving skills among all participants showed general improvement, as did their wellbeing, and their desire and willingness to learn. Enthusiasm increased, social skills showed improvement, and willingness and openness to try new experiences was increased.

During this work with children in special needs groups, the positive effect of VR upon pupils with developmental disorders cannot be overstated. For the first time these pupils find themselves in a world that makes sense to them. It is a world they can understand, and, crucially, it is a world they have a sufficient level of control over. This sense of agency is vital, and the importance of this to the children in question cannot be overstated. Furthermore, and while it is clear that further study is required, this feeling of increased agency seems to transcend the virtual world and have a positive effect on the real world too for these pupils, based upon their descriptions.

3.2 VR in Healthcare

This case study will look broadly at Viarama's work with senior citizens in nursing homes, hospices, respite centres, and hospitals. A formal study of our work in hospices has been undertaken by Queen Margaret University in Edinburgh in conjunction with St. Columba's Hospice, and this study will be published in late 2020. Anecdotally, we have collated hundreds of positive comments, reviews, and general feedback which overwhelmingly show the many benefits to not just to the individuals we work with, but also to their families. The benefits to family members are most keenly observed in our work in hospices, where family members are most often present. A notable example occurred during one of our first sessions in a hospice.

The lady concerned was in some significant pain prior to her session and on the day wasn't sure if she would be able to participate. The session was delivered at her bedside after the bed was wheeled into the room we were using. She was very frail at this point, and great care was taken to ensure her comfort at all times. Upon entering the virtual world she chose, which was an undersea visit, she eventually broke into a beautiful smile, which was deeply moving not just for her husband and son, but also for doctors as well as Viarama staff.

Following this, the lady decided she would like to travel to places she had visited and 'loved' during her life, starting with where she and her husband had their honeymoon, and taking in all of the places in the world she had loved to visit. For all concerned this was a deeply touching session that was unforgettable.

After the session had ended, the lady's son spoke to Viarama staff and, deeply moved, said he hadn't seen his mother smile for many months. He expressed deep gratitude at how beautiful it was to see her experience happiness one more time. There have been a great many examples of this kind of deeply moving response to Viarama's work in hospices, and this clearly shows the benefits of VR for people receiving palliative care, and their families.

Participants often have a profoundly emotional response to unexpectedly being able to visit parts of the world in VR that have had strong significance in their lives. This emotional response is something Viarama now includes in its briefing for all participants prior to their session; along with other psychological and practical explanations of what they are about to experience. This briefing has been carefully developed over the course of Viarama's work and has been refined to be as efficient as possible, whilst still conveying the essentials to ensure participants are safe, happy, and prepared for the experience.

In hospices, nursing homes, respite centres, and hospitals Viarama's work is predominantly, but not exclusively, done with senior citizens. The people worked with present with all levels of physical and cognitive ability. In the vast majority of sessions Viarama staff will guide participants through the experience by controlling the virtual world on their behalf, and this has proven to be the most effective way of conducting the sessions. In these settings, virtual travel is a very popular experience and from a menu of options this is by far the most selected by participants. Other experiences used include drawing, painting, and sculpting, driving, flight simulators, and various other games and applications as requested. It is worth noting however that on occasion, senior citizens just want to shoot zombies as much as the next person.

4 Feedback from End Users

The anecdotal feedback mentioned above has been one of the key indicators to date that Viarama's service is providing benefit to the people they work with. They have received hundreds of examples of this kind of feedback in public on social media channels, and in private in both sessions and the aforementioned formal research project which was participated in. Many participants are very keen to offer their feedback as they understand the requirement for more people to gain access to the service and see their public feedback as a way to achieve this.

One physically disabled user stated that he 'forgot about his wheelchair for thirty minutes' during his session, where he was weightless on space station.

One man receiving palliative care had devoted much of his life to his passion for motorcycling, and Viarama was able to let him enjoy a few more virtual motorcycle rides before he died. He stated to his family and carers that these experiences made his last few days much more enjoyable, and for a little while let him simply enjoy something, and not to worry about his situation.

A lady who had not been able to travel in her life due to her severe disability spoke of her deep desire to see Mount Everest. When Viarama staff took her to the top of the mountain and let her see the view, she cried tears of joy at being able to in some way realise her dream.

A lady who was receiving palliative care who loved fine art was able to visit the locations where some of her favourite paintings were created, as well as some of the finest art galleries in the world across a few sessions. She was also able to travel to some of the cities in the world she wanted to visit, including Paris, Amsterdam, and New York. Overcome with emotion, her gratitude was captured in a BBC report on our work, available online.

Phyllis was 101 years old when Viarama delivered her session in a hospice. Prior to our visit she had been feeling very down she said, as her contact with people was very limited. She was sceptical about VR but agreed to try anyway. Her wonderful life to that point had seen her travel the world as a child due to her father's career in the navy. She lived in many places but one in particular, Darjeeling, had stuck with her throughout her life, even though she had not been there since childhood, more than ninety years before her session. Viarama took her back to Darjeeling and let her travel through the area, and most astonishingly see her old home once again. Phyllis cried tears of joy to see her old home once again. Re-energised by the experience, she spoke at length about her childhood, her family, and her life to date. This session was fortunately captured by BBC Radio Four and some of it is available online.

Mick is a multiple sclerosis sufferer who had seen a sharp decline in his condition over the year prior to his session. He was unable to travel due to this, but was determined to put a brave face on it, and not let his condition get to him. In initial discussions with both he and his carers, Viarama staff became aware that he was deeply troubled by his condition, and they took great pains to explain the emotional component of the experience, and briefed him that sometimes an outpouring of emotion is to be expected. He was not the kind of man given to discussing his feelings however, and he curtly refused to acknowledge the possibility of him having an emotional response to this or any other experience publicly.

His favourite place in the world was the beach in Ibiza where his daughter lived, and he wanted to go there first. He explored the area for several minutes, awestruck at the experience in general, but specifically astonished to suddenly be able to virtually visit this special place and have it evoke such surprisingly strong feelings in him. Soon, he asked to stop and at this point he had what he later called 'a meltdown', which initially was a great source of embarrassment for him.

After the session ended, he spoke about his condition, for the first time, staff said. He went into detail about how his life had changed, how he could no longer do several of the things he loved to do, and how he felt about that. He was able to express sadness, anger, fear, bewilderment, and feelings of abandonment for the first time.

In his feedback he said he had been unable to confront his condition prior to this, and had largely been living in denial of it, but that the session has somehow acted as a catalyst to allow him to come to terms with his life as it now was. He spoke at length of his gratitude to have experienced something he regarded as profound in many ways, and expressed a strong desire to try the experience again, which later occurred.

There have been a great many more examples like those above in the work Viarama has done to date. VR in this context offers a profound experience that is unforgettable, deeply moving, impactful, energising, transporting, and highly beneficial to those they work with.

5 Future Outlook

Viarama is a social enterprise that to date has relied on private capital investment by its owner, alongside the revenue generated in an entirely new sector largely of its own making. The social enterpise has received no grant funding to date, and was recently turned down for the both of the two resilience funds it was able to apply to. Viarama's world-leading work is currently severely threatened by the outbreak of Covid-19 and there is a chance that the social enterprise may not be able to continue as all work has been forced to cease since March 2020.

This work is ground breaking, and its value is beyond question. If this work is to continue at all we must see a widespread willingness to embrace new technology in all of the sectors where we work. There is cause for optimism however, and the hope is that change that is long overdue will be delivered as a necessary response to Covid-19.

Viarama hopes to expand its work internationally in the coming year if at all possible, using VR to help as many schoolchildren and senior citizens as it possibly can in the process.

6 Conclusion

There can be no doubt that the VR service provided by Viarama is of significant benefit to those worked with in schools, nursing homes, hospices, respite centres, and hospitals.

As understanding of VR grows in the public consciousness it is hoped that demand for Viarama's services—an even mixture of both the technological and the psychological—will continue to increase.

It is inevitable that virtual reality has a significant role to play in both education and healthcare around the world both currently, and in future. It is Viarama's aim to deliver their VR service in every school, nursing home, hospice, respite centre, and hospital in Scotland. These are ambitious aims, clearly, and Viarama will not be able to achieve these aims alone. Support of all kinds is essential to allow the social enterprise to continue and hopefully to flourish in years to come.

Use of Virtual Reality for Hazard Safety Training to Reduce High Risk and Significant Safety Incidents and Increase Training Engagement

Anthony Jones

Abstract

Home safely, every day is a core value for LINX Cargo Care Group. Living this core value requires strong support from its people and an equal measure of innovation. In a first for the Australian supply chain and logistics industry, LINX Cargo Care Group (LINX CCG), has leveraged Virtual Reality (VR) technology to significantly lift positive engagement and safety outcomes. LINX CCG's award-winning VR safety training platform has not only cut the time required for training—while simultaneously improving the quality—it has also delivered an unparalleled level of consistency and notable savings to the Group in form of improved efficiencies.

Keywords

VR · Safety · Supply chain · Hazard reduction · Training · Logistics

1 Company Description

LINX Cargo Care Group (LINX CCG) is one of Australia and New Zealand's largest supply chain and logistics companies and brings together the capabilities of five market-leading operations built on more than 100 years of ports and logistics experience.

Together, LINX (Logistics, Port Services and Rail), Autocare Services, C3, Pedersen Group and GeelongPort employ approximately 3,300 people across sites in Australia and New Zealand.

A. Jones (✉)
LINX Cargo Care Group Pty Ltd, Sydney, Australia
e-mail: communications@linxcc.com.au

T. Jung, J. Dalton (eds.), *XR Case Studies*, Management for Professionals,
https://doi.org/10.1007/978-3-030-72781-9_10

LINX CCG customers come from a diverse range of industries right across Australia and New Zealand, from agriculture, oil and gas, aluminium and automotive to marine, mining and resources, food and beverage, and forestry.

Every year, LINX Cargo Care Group handles approximately more than 22 million tonnes of bulk products, almost two million tonnes of steel and more than 15 million tonnes of forestry products. LINX CCG also transports hundreds of thousands of new vehicles across Australia from shore to door and processes more than 400,000 vehicles for many of Australia's largest fleet operators.

The Group's vision is *"to deliver a connected supply chain, one move at a time"*. With an extensive footprint across Australia and New Zealand, their regional businesses play a key role in realising their vision for the supply chain and logistics industry, their customers, employees, and all key stakeholders. Owned by global asset manager Brookfield Asset Management and its investment partners (GIC, BCi, QIA), LINX Cargo Care Group has the regional capabilities and global connections to develop holistic, well-resourced logistics solutions.

Globally, Brookfield has more than US$575 billion of assets under management in infrastructure, property and renewable energy, across more than 30 countries globally.

2 Project Summary

According to Statista, the estimated market size of the global workforce training industry reached $370.3 billion US in 2019 (Mazareanu, 2020). However, Accenture highlights that many of these sessions are delivered in traditional formats like classroom-based seminars or online training modules. "While passive learning and memorization has been the past model, today's workforce requires a more active and ongoing approach to training in which employees learn through practical experience", a report from Accenture states (Raghavan and Rao, 2018). Further, a recent study has suggested that employee safety could be improved through use of Virtual Reality (VR) in Health and Safety training, such as fire evacuation drills (University of Nottingham, 2019).

Given their dispersed and diverse workforce, some of whom operate 24/7 in hazardous environments with large machinery, it is important to LINX CCG and its people that safety is continuously improved to create a compelling, simulated experience that cuts through and has an impact. Inconsistency in hazard perception, safety leadership and quality of training has been a significant challenge for the Group. The Group was undertaking regular safety training for staff, but it was often done on-site with potential hazards present and at the cost to the business of having the site shut down for training purposes. The courses would take a full day, with full pay for those employees attending. Further, there were site costs, insurance to cover and a myriad of other expenses to account for. To counter these downsides to training LINX CCG began looking at all available training options, and ultimately came across the possibility of developing or purchasing a virtual reality training module for its next wave of staff training.

3 Project Details

In a first for the Australian supply chain and logistics industry, LINX CCG released its VR safety training platform in March 2019.

LINX CCG worked with Australian communication, content and technology company Curiious to develop the VR program to help in "sending its circa 3,300 people home safely every day, across sites in Australia and New Zealand", in line with LINX CCG's mission. Further, LINX CCG saw the potential for the VR/AR solution "to transform the safety training process, reduce accidents and increase engagement within the organisation" (Fig. 1).

A six-minute training simulation was developed to safely expose its employees to workplace risks, and to increase recall and proficiency from training sessions. LINX CCG's LINX Port Services operation at Port Kembla, south of Sydney, in the Australian state of New South Wales (NSW), was chosen as the pilot site for the first iteration of the VR training. Port Kembla is NSW's largest motor vehicle import hub and grain export terminal and is the second largest coal export port in the state. The port handles a range of dry bulk, bulk liquid, and general cargo. LINX loads steel aboard ships at Port Kembla, and needs preparation and discipline within risk zones, with the safety exclusion zone around forklifts and heavy loading critical (more information can be found at https://www.nswports.com.au/ports-and-facilities/port-kembla/).

Fig. 1 Members of the LINX Cargo Care Group team using the Curiious virtual reality safety training system

Fig. 2 Staff appear as avatars in the virtual environment

Part of the platform deployed by LINX CCG was the Curiious IQ platform, which allowed trainers to annotate and teach live in the virtual space for up to 50 trainees at one time, all represented by an avatar within the virtual space. This allowed for the training to be monitored by the trainer in real-time for immediate feedback and subsequent instruction, and to provide those trainers with the ability to also monitor behavioural patterns such as reactions and stress levels.

LINX CCG decided on groups of up to eight trainees for its initial sessions. The initial VR training session opened within the virtual space to a 3D representation of the Port Kembla site office, followed by an interactive task to familiarise users with the controls and with the safety information displayed in the site office. The "*Immersion—360 Video*" and subsequent "*Interaction—Site Planning*" activity and assessment included lessons on the safe loading of heavy vehicles from a ship to the back of trucks. The program was able to visualise 12 angles of an activity such as a forklift carrying a load, and the surrounding layout (Fig. 2).

Choices made in the "*Interaction—Site Planning*" session could be recorded for playback purposes, while the trainer was also able to witness choices made in the virtual space in real-time via the IQ platform and coach them live during each session.

Fig. 3 Staff can train in a lifelike VR environment captured using 360-degree cameras on site

As a result of the implementation of the program, LINX CCG was able to shorten the training window from a full day to under two hours. Using the simulation, staff recall of the session lessons were greater and proficiency improved as the Group was able to run through the sessions with its attending staff simultaneously and repeatedly.

Staff could attend two hours of training instead of eight and could work immediately afterwards. Lessons could be repeated later, with the headsets already purchased and operational, and at no additional cost to the company.

In the six months since the first VR training module, there had been a 71 per cent improvement in hazards reported, and a 90 per cent improvement in near misses reported. Additionally, High Risk and Significant Safety Incidents were down 50 per cent, and recordable workplace injuries were down 10 per cent (Fig. 3).

4 Feedback from End Users

John Hall, LINX CCG's Business Manager, Port Kembla, said the Curiious IQ platform has transformed the safety training process and how it is delivered for the Group.

"Traditionally, we'd use PowerPoint presentations [and] printed documents, and the first thing a person would do is flick through and see how many pages there are, and then we'd lose [their interest] straight away," said Hall. "The employees have loved it. It's probably the first time I've ever seen people interested in training in 20 years."

In the first six months of the program being used, safety engagements conducted increased 32 per cent while training time had been reduced by up to 70 per cent.

LINX CCG's Executive General Manager—Health, Safety and Environment, Peter Seaman, said people had become more engaged with training and "wanted to train because of the VR environment".

Fig. 4 Staff at Port Kembla are introduced to the Curiious training platform and VR headsets

Michael Kemp, Systems & Governance Manager—Health, Safety and Environment for LINX CCG, said the program had demonstrated a measurable impact on safety outcomes.

"The appeal of this unique virtual reality safety training is the fact that our workers can be exposed to high-risk situations in a 100 per cent safe environment," he said.

"In addition, the VR training solution is portable which enables a consistent training experience and will ensure we can roll out this unique solution at our sites across Australia and New Zealand. This addresses a key business challenge to ensure employees are trained the same way across our organisation."

The project was recognised with the Supply Chain Innovation & Technology Award at the 24th annual Australian DCN Shipping & Maritime Industry Awards held in Melbourne in 2019 (more information can be found at https://www.linxcc.com.au/news/national-shipping-and-maritime-industry-awards-success-for-linx-cargo-care-group/).

LINX CCG also received a 'Highly Commended' award from The International Cargo Handling Coordination Association (ICHCA) at the organisation's ICHCA 20/20 Cargo Vision international conference in Malta in November 2019 (more information can be found at https://www.ichca.com/3rd-tt-club-innovation-in-safety-award-presented-to-kunz-csxit-and-yardeye) (Fig. 4).

5 Future Outlook/Roadmap

LINX CCG will make the VR training sessions available to all sites in Australia.

Further, due to the reduction in on-site work-related injuries and injuries since the program has been in use, LINX CCG is also looking at additional projects within its Group for which VR training could prove beneficial, and not just for safety training.

6 Conclusion

The use of the VR training program has delivered measurable benefits to the safety on site, as well as lowered the overall cost of training by reducing the on-site disruption caused through initiating and undertaking live training sessions, thus delivering a return on its investment into the platform development and required headsets.

LINX CCG will continue to use the current program for training purposes and will also look to roll out the platform across its sites and businesses following the quick and successful integration of the technology into its training modules.

With end-user feedback unanimously in favour of the training modules, LINX Cargo Care Group believes the rollout of the technology across sites will deliver further benefits and measurable increases in health and safety standards.

References

Mazareanu, E. (2020). *Spending in the global workplace training industry 2007–2019,* Statista.

Raghavan, R., & Rao, P. (2018). *Immersive learning for the future workforce,* Accenture.

University of Nottingham. *Virtual reality training could improve employee safety.* ScienceDaily. 16 September 2019.

Stations AR Simulated Digital Twin

Darren Ackers

Abstract

PAULEY developed a highly innovative platform for augmented reality visualisation and data integration to create a digital twin of HS2's Old Oak Common Station. Immersive, mixed reality technology aims to help revolutionise the way in which rail station staff are trained and upskilled. Using interactive overlays, provides a platform for enhanced remote training, business continuity and incident management.

Keywords

Augmented reality · Simulated · Digital Twin · Rail · Training · Station operations · Digital engineer · Innovation · Immersive · Visualisation · Technology

1 Company Description

PAULEY are a UK based multi award-winning team of immersive technology specialists who help deliver innovative and effective digital solutions for stakeholder engagement, maintenance/fault finding, project management, skills/training and competency management.

A pioneer in establishing the Digital Engineer of the Future, PAULEY analyse, design and develop innovative operational performance support and agile blended learning applications using a range of digital immersive technologies including Augmented Reality (AR), Virtual Reality (VR), Mixed Reality (MR), Artificial Intelligence (AI) and Internet of Things (IoT). PAULEY provide a range of Tier

D. Ackers (✉)
PAULEY, Milton Keynes, UK
e-mail: darren.ackers@pauley.co.uk

T. Jung, J. Dalton (eds.), *XR Case Studies*, Management for Professionals,
https://doi.org/10.1007/978-3-030-72781-9_11

1 businesses and clients with bespoke agile learning, upskilling, train the trainer and competency management solutions. PAULEY are world leaders in spatial computing applications for industries who help clients to optimise operational performance, reduce risk, and increase safety and customer satisfaction through pioneering applications of immersive technology.

2 Project Summary

A simulated digital twin was produced of the new High Speed 2 (HS2) project in the United Kingdom planned to open in 2026. HS2 is the largest infrastructure project in Europe and is a £100bn state-of-the-art, high-speed line linking up London, the Midlands and the North, serving eight of Britain's 10 largest cities. The Department for Transport (DfT) Innovate UK funding competition focused on the development of the HS2 Stations augmented reality digital twin application, which demonstrated a range of learning applications for a scenario based digital twin. PAULEY lead the project with HS2, National College for Advanced Transport & Infrastructure (NCATI) and WSP, working together to deliver the new multipurpose synthetic digital environment.

The aim of the project was to demonstrate for the first time, the unique capabilities of an AR digital twin platform within the railway station context and to show how future rail station staff could be trained and upskilled using AR technology. The AR platform allows the station operations team to collaborate in this new synthetic environment to simulate the various systems and processes typically found in a stations operations control room. Various features were incorporated into the application that covered the operation of the station, dealing with information fed from service operations, management of customer through various safety and security systems, and communication with staff and passengers to ensure the station operates smoothly. The inclusion of signage and wayfinding information was included to help train various staff on station familiarisation. The basic management of assets allowed an understanding of how these affect the operation of the station.

A number of scenarios were created to allow the user to practice dealing with disruptive situations and train for effective crowd management and communication.

3 Project Details

PAULEY's vision of the Digital Engineer of the Future lead the development of a highly innovative platform for mixed reality visualisation and data integration to create a digital twin of the largest of the four HS2 stations, Old Oak Common. Immersive, mixed reality technology merges the real and virtual worlds, and using interactive overlays, provides a platform for enhanced cooperation, team coordination and real time analysis.

The technology provides a valuable ability to devise and enact synthetic learning and continuity-management scenarios for future emergencies, maintenance, fault finding and impact planning. Augmented Reality aligned to HS2's strategy of using cutting edge technology to ensure end users were comfortable with using headset based spatial computing which would be mainstream when stations become operational. The outputs will serve to minimise future disruption for the travelling public, many years before the physical station is constructed. Stakeholders, employees and customers are all able to take the path through the twin environment that best reflects their personal priorities. The environment incorporates 'customer personas' to ensure operating staff are fully aware of the difficulties that these individuals may encounter and can take active measures to provide assistance as necessary.

The timing of this project was advantageous as it coincided with the commencement of station development. WSP, the contracted designer, provided expert assistance into the functional view and imagined environment, providing a true simulation of the station to ensure it can be future proofed and managed.

The simulated digital twin facilitated delivery of training and competency in operations (control room, platform, maintenance, etc.) and customer management, improving the skill and competency of staff, boosting their ability to meet and exceed customer expectations. The findings have prompted improvements to plans for signage and customer information as well as physical remodelling to optimise retail return-on-investment and the customer experience.

PAULEY and the project partners delivered a demonstrator structure for the HS2 super hub such that it can be scaled and rolled out to other HS2 stations and applied retrospectively to other major stations across the network (e.g. station modernisation or refurbishment projects).

The AR platform allows trainees to understand how the station is operated and to ensure the station runs smoothly by incorporating the following features:

- Station Operations—Manage deployment and locations of platform staff to ensure the train dispatch process runs smoothly. Use the safety and security systems to manage issues and incidents.
- Service Operations—Manage information fed from a simulated Network Integrated Control Centre (NICC). The data will show the augmented reality trains arrive, dwell and depart the platforms, position of trains and status of signals on the track diagram. Customer information will also be fed to the user to identify issues with other transport links or key events and incidents.
- Customer Management—Observe pedestrian flow using 'heat' maps. Manage crowd management facilities such as barriers and gate control. Access to the CCTV system to view up to 8 cameras within the system.
- Communication Systems—Manage Interfaces with staff through voice or messaging to provide service information updates. Make announcements for emergency situations (regular platform announcements are automated). Update digital display screens around the station.
- Signage and Wayfinding—Locate station and platform assets, view the platform zoning area system.
- Asset Management—Manage station assets and facilities such as lifts and escalators using a simulated monitoring system.

- Scenario-based Training—Trainees can prepare for future maintenance and operation scenarios allowing them to first practice within the digital twin environment.

The AR platform provides a mechanism for stakeholders to collaborate in a new digital environment rather than playing out the consequences in a physical station at the potential inconvenience to the travelling public.

The way data is often shared and communicated can be unstructured, patchy, confusing, often contradictory and rarely available at the point of need. The new platform unites individual and business requirements in a common data environment that utilises the advantages of mixed reality technology and enhances the customer journey at the station. Mixed reality blends the physical and digital worlds. It is the next evolution in human, computer, and environment interaction and unlocks possibilities that before now were restricted to pure imagination.

The core platform enables stakeholders, employees and customers to undertake a complete digital journey of the station environment in a safe and collaborative real time augmented space. The innovative platform:

- Updates the station development as the project progresses through subsequent build stages. The conceptual information will be replaced by actual design and construction data.
- Integrates with the internet of things data feeds from station systems and products.
- Is accessible from a real-time collaboration point of view in AR to gather information.
- Delivers training and competency in operations (control room, platform, maintenance, etc.) and customer management. Improves the skill and competency of staff, and their ability to meet and exceed customer expectations.
- Incorporates customer personas including Equality Act protected characteristics (e.g. the elderly, individuals travelling with children and/or luggage, wheelchair users, blind, deaf/hard of hearing and cognitive impaired) to ensure that operating staff are fully aware of the difficulties that these individuals may encounter and can take active measures to provide assistance as necessary.
- Provides the ability to scenario plan, particularly for maintenance, fault finding and the consequential impact planning to minimise actual disruption for the travelling public. Potentially including an improved customer information feed to manage expectations and inform on corrective actions being taken.
- Enables asset information modelling and building information modelling to be undertaken from a place of safety. The use of AR removes people from a position of danger and increases the quality of communication.
- Provides a collaborative environment in which other stakeholders can visualise their concepts within a station environment prior to development to ascertain impact and RoI.

Simulated AR digital twin being demonstrated to Mark Thurston, CEO, HS2

The project was led by PAULEY, but in essence was a true collaboration between all the project partners—PAULEY, HS2, NCATI and WSP. By working closely with all project partners, a platform was delivered that is multipurpose and works across industry and sectors.

The platform will drive efficiencies within HS2 and beyond, particularly through the delivery of more cost effective training and by ensuring that processes and procedures are undertaken in a more time efficient way, driving further cost savings.

PAULEY is building upon the digital twin concept and introducing more intricate and elaborate features—the HS2 digital twin station will augment the physical and virtual worlds.

The project deliverables were truly immersive and went beyond the limitations of AR. PAULEY is one of only 50 European companies to be accepted onto the prestigious Microsoft Mixed Reality Partner Program, an integrated program focused on enabling and supporting digital agencies, systems integrators, and solution providers who are committed to building mixed reality solutions for their clients.

HoloLens provides a self-contained, holographic computer, enabling the users to engage with bespoke digital content and interact with holograms in an immersive environment. The mixed reality environment combines elements of both AR and VR, real-world and digital interactive objects.

AR offers a seismic shift in working arrangements, data collaboration and support between rail technical staff, operations staff and the supply chain. The real benefit is that through this data environment, any physical point can be made available to the customers for real time management and communication. This presents a significant cultural and behavioural shift in communication. Simulation can be incorporated with actual live feed data for immediate performance feedback to deliver a superior solution, providing an enhanced customer experience.

Currently data is large, complex and can at times be somewhat chaotic. This innovative project introduced a format, structure and a mechanism for viewing the data in a collaborative way.

The HS2 simulated digital twin will, eventually, be connected and incorporate live stream data regarding the current status. It will not just be a piece of CAD software with a model in it representing what is going on; it will derive value from data feeds and present this information via an augmented experience. The vast amounts of data collated from the station's assets will prove to be invaluable to different stakeholders in a multitude of ways (maintenance, fault finding, wayfinding, stakeholder engagement, piloting new innovations, training etc.). PAULEY helps organisations collaborate and empower their teams to drive a more efficient and productive network and workforce through immersive digital experiences.

4 Feedback from End Users

End users were fully engaged throughout the development of the digital twin. The first round included a review session with the HS2 Infrastructure & Maintenance Team.

> There are clear and wide-ranging benefits of this technology being applied to both stations and the linear asset from a maintenance and asset management perspective. Everyone in the room was keen to see this project extended to include Rapid Response (and support for field technicians by those in maintenance depots and NICC), training, competence and familiarisation, defect rectification and reference to technical documentation (such as operation and maintenance manuals and company technical standards) during planned maintenance. Additionally, we see benefit in simulating complicated possessions and isolations, particularly using on-track machines, to improve safety of workers while on the track.
>
> Simon Morley, Head of Infrastructure Maintenance, HS2 Ltd

We recruited 8 users to the specification co-created with HS2. These users came from a range of current operational station and management positions to provide insight into the usability, accessibility and training value of the tool. The NCATI provided 2 of their students to attend a session, in part to provide the view of someone who hasn't stepped into a real control room of an operational station and may have different attitudes to technology.

Some reactions from users, focusing on the design of the tool.

Users stated they were impressed by the technology and the abilities it afforded them. *"Initially I was just gobsmacked by what you could do, where to go—there's that many options."*

> I was surprised by how realistic it looked.

Users liked being able to see a model of the whole station, and the ability to switch between floor levels, enabling them to extract information easily. *"I like that I can have a quick view of any part of the relatively massive place straight away. I feel free to choose what piece of information I want to get and jump from one to another."*

Users could see the benefit of using the tool prior to starting a new role, allowing them to familiarise themselves with the environment. *"It would help a lot...then people could create an idea of the station before they actually go on the floor."*

> In theory you'll have seen pictures in front of your eyes. So, when you actually walk in, you'll be thinking, 'oh yeah, I've seen this'.

Users could also see the opportunity for uses beyond the scenario they experienced, such as training for evacuations.

Many users were impressed by the visual style of the tool, including the icons, the line map, the control panels (while the CCTV was difficult) and the station model. *"It is definitely going to be very useful for all those people who'll potentially be trained in the future. Especially for new facilities, new information, new knowledge."*

> It would be very useful for training people from different stations who had never actually been physically in the place... the device could be used to show the entire station from sitting in a room, allowing them to get familiar with the geography and infrastructure of the place.

Many users noted they would see the tool as a safe environment to practice what they were learning as they grew confidence in their roles. *"It's easier to go into it knowing that, if you make a mistake, it's just as simple as turning it off and restarting it. Whereas if you went into a real station environment, and you got something wrong, it is a lot more stress. With this, you have the room to make mistakes and you gain confidence along with it."*

There's space for integration between digital and on-site training, with learning on the floor picking up directly from where the device leaves users. This is especially helpful as the stations are still being built. *"Having it as a training package as a whole would be very stimulating, because you've got the hands-on training on one hand and the visual augmented reality on the other."*

5 Future Outlook/Roadmap

Following on from the simulated digital twin the next natural step is to integrate it into a true digital twin. Once physical sensors and automated systems are commissioned at Old Oak Common the existing digital twin model will be integrated with the internet of things data feeds. As a result of this process, the digital twin will mature with real world assets and deliver added value over a similar timeframe for mechanical and electrical systems and construction stages thereafter. It is anticipated that project managers and asset managers will return to the digital twin to finesse plans for future refurbishment and upgrade projects.

As this functionality is finalised and goes live, the learning and productivity applications for the digital twin model will grow significantly. Real-time operations data will become accessible in AR so future managers and technicians will be able to identify challenges or faults and collaborate to rapidly find solutions that optimise productivity. This new capability is likely to influence future working practice at major train stations. The potential to provide staff with headsets to virtually view stations may reduce the need for control rooms at expensive, central locations.

Regular engagement with scenarios will allow trainees, maintainers and operators of train and track to explore offline digital twins that exhibit typical or historic operational data, or a real-time digital twin showing live data. This will allow them to develop the knowledge, skills and confidence they need to support effective operation.

Seamless collaboration between trainees, maintainers and operators will ensure they are all fully equipped to respond to any given situation during day-to-day operations. This will improve the running of trains, stations and track, thereby fundamentally improving infrastructure productivity and the customer experience.

Central to successful commercialisation is the development of a product that meets end user requirements. An understanding of customer pain points is a pre-requisite for exploitation—this must be supported by the creation of technologically innovative solutions to address the identified issues.

The digital twin platform will form part of a commercial offering for PAULEY. The aspiration is that this project will result in a commercial offering which will be offered to rail management companies to map and manage their existing stations, plan refurbishment, and construct new stations. Beyond the confines of the rail sector, opportunities for this platform technology also exist within large transport hubs (airports, seaports, etc.), large shopping centres, event spaces, and more. The transport and infrastructure supply and value chain includes passenger vehicle manufacturers, infrastructure manufacturers, building construction, power supply and retail.

6 Conclusion

The Rail Sector Deal looks to build on the strong partnerships between the rail sector and the government to exploit the opportunities of new digital technologies. This will improve the efficient use of the GB rail network capacity and enhance the experience of the passengers by improving the service they receive.

Digital twins help organisations collaborate and empower their teams to drive a more efficient and productive workforce through immersive digital experiences. Clients including the National Training Academy for Rail (NTAR), Siemens and NCATI have already experienced savings of 25–30% in training costs and a significant reduction in training asset costs when using a digital twin as opposed to physical training options.

US Air Force Weather Training Platform: Use of Virtual Reality to Reduce Training and Equipment Maintenance Costs Whilst Improving Operational Efficiency and Retention of US Air Force Personnel

Ramzy Ross, Dainius Slavinskas, and Eric Mazzacone

1 Company Description

Myriad Global Media was founded in 1989 to develop high impact communications and training for private and public sector organisations. Over that period, Myriad Global (MG) has worked across multiple industry sectors including: Energy, Transport, Defence, Engineering, Healthcare and Education with clients as diverse as the Cabinet Office, Disney and ExxonMobil. It is headquartered in London (UK) with regional hub offices in Abu Dhabi (UAE) and Houston (Texas, USA), and a professional support network in over 30 further countries worldwide. The Founder and Group CEO is James Thompson; Dr. Ramzy Ross is Head of Innovation and Dainius Slavinskas is Head of Technology.

The Innovation and Technology Solutions team are a key engine of growth for MG and are focused on research and development. They apply an approach that combines strategy, user insight and technological engineering to create bespoke solutions that deliver a return-on-investment. Key capability areas include immersive and interactive technologies involving Mixed Reality (MR), and associated digitalization and deployment, of Augmented Reality (AR), Virtual Reality (VR) and associated applications.

2 Project Summary

The US Air Force (USAF) is required to operate in varied, and often challenging, environments. The US Air Force recruits are continually being trained to become highly skilled practitioners who can readily adapt to any potential changing

R. Ross · D. Slavinskas · E. Mazzacone (✉)
Department of Innovation and Technology Solutions, Myriad Global Media, London, UK
e-mail: ramzy.ross@myriadglobalmedia.com; dainius.slavinkas@myriadglobalmedia.com; eric.mazzacone@myriadglobalmedia.com

T. Jung, J. Dalton (eds.), *XR Case Studies*, Management for Professionals,
https://doi.org/10.1007/978-3-030-72781-9_12

situation. The Air Force Agency for Modelling and Simulation (AFAMS) was established in 1996 and is tasked with enhancing, and leveraging, modelling and simulation 'to support and facilitate integrated, realistic and efficient operational training across war-fighter domains to enhance full-spectrum readiness' (further information can be found on https://www.afams.af.mil/About-Us/Fact-Sheets/Display/Article/429786/afams/ [Jan. 7, 2019]).

The US Air Force has been increasingly looking to immersive and synthetic solutions to support the development and learning of its teams (e.g. Bell et al., 2020). This particular use case involved the creation of a VR solution to train recruits in setting up, and operating, the TMQ-53 Weather Station and Broadband Global Area Network (BGAN) equipment. Such systems are portable and automatically take observations, in up to one-minute intervals, enabling flying missions across the globe. The data produced can be utilized by a weather observer in the field or by the Air Force Weather community using satellite communications.

A solution was needed to ensure that 'real-world' USAF technical training objectives could be replicated with the precision required for operational effectiveness. The use case emerged from the logistical challenges USAF was experiencing in not having enough readily available equipment to balance the requirements for individual training units and for live missions. A digital twin VR solution was seen as desirable because it would enable the capability to train individuals, at any location, without having to potentially disrupt live mission activity. For interest, further information on the digital twin approach, and related developments, has been previously covered in a recent publication (Fuller et al., 2020). The solution would also assist in reducing equipment related costs as related training would become less reliant on more costly, real-life equipment as a result of incorporating the VR approach. It was also seen as a training platform that would be embraced by trainees that came from a generation that were familiar with gaming technology and saw the implementation of immersive technology as a positive development in their training cycle.

3 Project Details

3.1 Getting Started: Aligning Hardware Specifications and Requirements

The MG development team was tasked to deliver a standalone, locally deployable VR application to enable immersive training of USAF trainees on the digital twin equivalent of the Vaisala TMQ-53 Weather Station and Hughes' BGAN. Figure 1 shows an image of a similar portable real-life weather station set up.

The US Air Force first Weather Group procured first edition HTC VIVE (Taoyuan City, Taiwan) VR headsets and Windows-enabled computers (Microsoft Corporation, Washington, USA) with GeForce GTX1060 graphics cards (Nvidia, California, USA). This set the baseline for the development, and design, of the virtual reality training application. From the outset the MG development team were

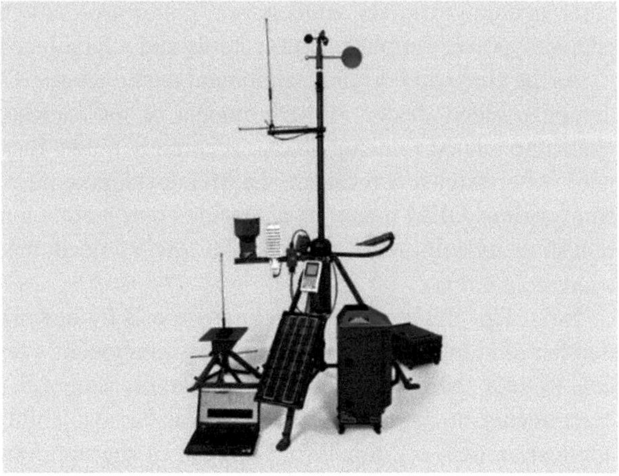

Fig. 1 An example of a Vaisala TacMet® Tactical Meteorological Observation System used to develop a digital twin equivalent for the USAF VR training application

aligned with USAF hardware specifications, and related requirements, to ensure the application would run without any performance issues whilst also aiming to achieve the highest levels of digital twin fidelity and asset quality. Unity 3D (California, USA) was used to develop the application. Autodesk Maya (California, USA) and Substance 3D (California, USA) was used to create visual 3D elements and objects.

The MG development team was tasked to deliver a standalone, locally deployable application with the following deliverable items defined:

- Real-time and hands-on experience for recruits to be able to setup, operate, teardown and troubleshoot the TMQ-53 weather station equipment.
- Mimic three weather conditions (sun, rain, and snow) that will require the user to think differently and ensure they adapt their methodology to complete the setup task effectively.
- Trainer-trainee multiuser mode.
- Training and assessment mode.
- Three different backdrops/locations, each with a different level of difficulty (area surrounded by buildings, area surrounded by the trees and a location in an open airbase).
- Tracking and collection of user actions and time taken to complete each assembly step.

3.2 Developmental Stages and Challenges

The specific weather station model was the TMQ-53—an off the shelf system manufactured by VAISALA (Vantaa, Finland) and used by USAF (Roles, 2017). The kit comes packed in five durable cases and is shipped to missions with some additional items, such as sandbags, to weigh the deployed equipment down for enhanced stability purposes. There are approximately 80 individual component parts including: tripods, wind sensor, rain sensor, antennas, modems, cables, power adapters, power supplies, various tools and a laptop computer.

As the client and original equipment manufacturer (OEM) were unable to provide computer-aided design (CAD) models of the various components, a key initial challenge related to being able to accurately model digital versions of each component. After extensive research, the MG development team collated reference images from various OEM materials alongside coordinating a procedural video demonstration showing complete setup, and teardown, procedures as demonstrated by a USAF team at an airbase.

Next steps involved the development of a framework that could handle the large number of individual components, each component's respective snap targets, and be able to track the correct equipment assembly sequence. Due to the large number of components, an additional challenge was the availability of testing time to meet the application delivery timeline. For each of the approximately 80 component parts there was a need to test for potential bugs/errors, carry out any subsequent bug fixes, correct procedural sequences and allocate time for the addition of potential new features. All had to be considered and carefully managed.

Finally, consideration had to be given towards security. In particular, restricted access to the airbase network resulted in MG being unable to carry out any of the application deployment work. Further, due to the same network infrastructure-related restrictions, the MG proposed cloud-based multiplayer Photon Engine (Hamburg, Germany) could also not be deployed. As a result, the multi-user feature had to be subsequently removed from the scope-related deliverable items.

3.3 Training Space Set Up

The training application was designed to work with room scale user tracking allowing the user free movement within the constrained space limited by the VR headset cable length. To avoid cables tangling and being damaged, a retractable cable management system was assembled, and fixed to the ceiling, at given locations. HTC VIVE base stations were placed around the training zone perimeter to establish VR headset and hand controller tracking within the virtual space. Figure 2 provides an illustration of the training zone setup and perimeter.

3.4 Virtual Reality User Experience

The devised training application formed a blend of the VR application and Computer Based Training (CBT). Some elements, such as data entry or monitoring, took place on the workstation monitor whilst training took place via the VR environment.

The training experience starts with the trainee, or instructor, entering personal identification details on the respective PC terminal. The user is then prompted to select a 'User Mode' by selecting either 'Trainer' or 'Trainee' mode and this is then followed by an option to select either 'Tutorial' or 'Assessment' mode. Following

Fig. 2 A USAF recruit wearing a VR headset and interacting with virtual objects with handheld wireless controllers

this, the user is prompted to select the specific weather conditions the environment should encompass—either 'Sunny', 'Rainy' or 'Snowy'. The user is then instructed to put on the VR headset. User placement within the 3D environment is randomised but the user has the ability to change the location. Assessment of the environment is via a virtual action control user interface (UI) triggered with a hand gesture that is captured by the handheld controller. Additional actions can be triggered as Fig. 3 illustrates below.

The training mode begins with the user opening storage crates and taking out the required parts of the equipment. Each component has information tooltips identifying what part it is and where it connects—an example is shown in Fig. 4.

To continually guide users and aid the training experience, when a part/component is picked up, a hologram appears on the snap target which guides the user to the correct placement of the part (Fig. 5). The hologram aids and tooltips are removed if the user is in the assessment mode and all connections and part placements must be completed, in the right sequence, without any aid. To pick up objects, the user must press and hold the controller trigger button which is named 'one button for all'. The same trigger button, and functionality, allows for the securing of bolts (Fig. 6), placement of sandbags (Fig. 6), extending parts, and collapsing parts amongst the

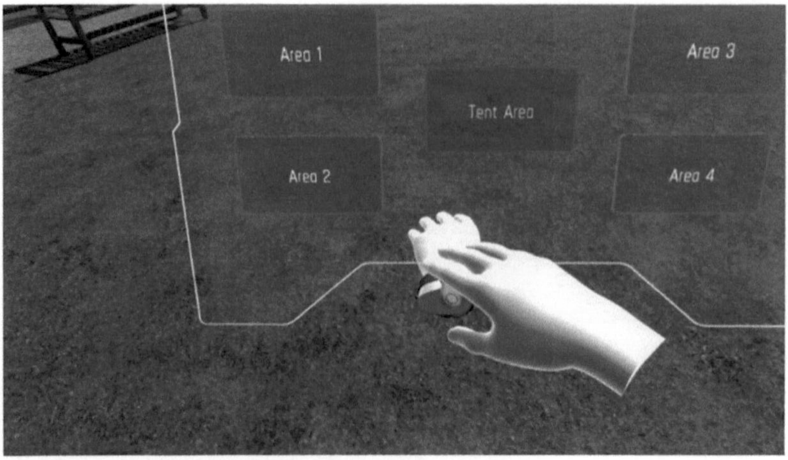

Fig. 3 Hand gesture recognition triggers the action control UI. In this example, the user swipes the right-hand controller over the left-hand wrist to open the UI

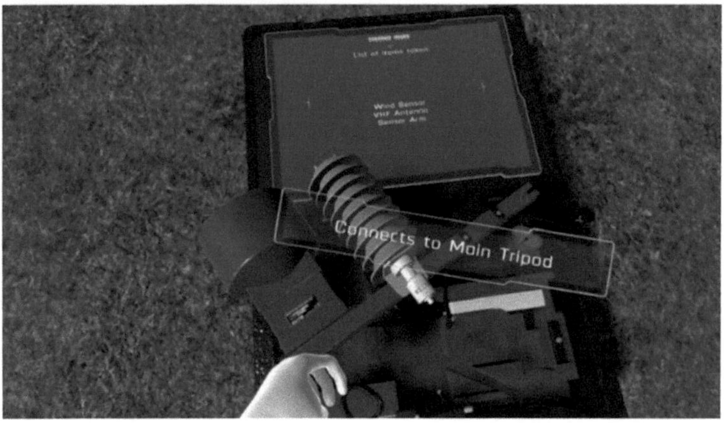

Fig. 4 Labelled equipment parts help the user to identify and connect each respective component to the right place

other actions required in order to successfully complete all steps of the training application. This approach simplifies user-hardware interaction by taking away time needed to learn how to use the controllers, especially for recruits who have never used VR systems and controllers prior.

The MG development team used procedural cables to visualize equipment cable connections. The user would pick up a plug, already having identified the correct markings and pin count, and plug it into the right socket. The cable is then procedurally traced from one connection point to another. When a connection is made correctly, a visual indicator (LED light) turns on to indicate power or data connectivity. In order to replicate real life scenarios, and train recruits to deal with

Fig. 5 Hologram identifying the precise part/component-specific snap target

Fig. 6 Trainees have to make suitable judgements for the placement of the main tripod and to ensure the equipment is firmly in place with actions including securing with bolts and placements of sandbags

varying conditions, a selection of power adaptors and extension blocks were introduced. As an example, there may be a situation where the given location power outlet has a UK 3 pin socket, but power supply would be provided with a US 3 pin cable, thus, users are challenged to choose the correct adaptor to complete the connection. Users are also asked to consider an appropriate setup location—for example, not to place the equipment too close to a building, or other obstacles, that might interfere with the station's accuracy.

This approach was also extended to train users on making the right decisions depending on weather conditions. For example, if weather conditions were rainy or snowy, user should use provided concrete blocks to support the power supply off the ground and, if this step is missed, the equipment will report a fault and warn the user accordingly.

Audio effects were also introduced to enhance realism and the user experience. Each action would have a different sound loop to give the user audio-based feedback that an action has been applied. Unexpected audio sounds were also introduced to create better audio realism and 'easter eggs', as named by MG developers, were also featured to further enhance the user experience with noises such as those of a jet flying by randomly.

Upon the successful setting up of the weather station the user progresses to the next phase of the training which involves a similar approach but applied to the BGAN equipment.

4 Feedback from End-Users

4.1 Continual Feedback and Responsiveness

It was imperative that feedback was continually obtained throughout the training application development process to fine tune the approach and enhance the user experience. As part of this process, the MG development team released Alpha and Beta versions of the application to keep recruits engaged whilst also enabling the continual testing of the application at early stages of development. Further, this process also allowed for early engagements with recruits, many of which had never used a VR system prior, and demonstration sessions were pivotal for familiarisation purposes. Various forms of feedback were received throughout the development process—from the inception phase to delivery of the finalised application for training purposes.

4.2 Using VR as the Solution

Virtual reality systems can often involve a big learning curve, especially for new users. The Myriad Global approach resulted in recruits being able to effectively navigate and interact within the virtual space, alongside carrying out the required equipment assembly, in a time appropriate manner. Providing a VR-focused solution also enabled reductions in costs associated with the purchasing of real-life equipment, and associated maintenance costs, which was also a key deliverable. Tech Sgt, Rob Thomas, first Weather Wing Systems and Training Officer for USAF, provided the following related statements:

> VR cuts down on the need to purchase additional equipment, eliminates temporary duty travel and dramatically lowers the potential for equipment breakage. The real savings come from Airmen not being taken away from their missions.

> Saving costs is what everyone wants to hear, but opportunity cost is today's enemy, particularly time. A trainee can be immersed and familiarize him or herself without ever seeing the equipment and a member already certified can use it towards refresher training.

The solution either simplified, or removed, related logistical considerations relating to equipment availability across USAF sites. It was acknowledged that the

VR solution would not completely replace related training activities but would be focused on providing familiarisation as well refresher training. After 6 months of deployment, over two million USD in related costs were saved by USAF.

In terms of hardware, no major related challenges were experienced or reported. However, a number of considerations were raised for future developments of similar solutions. Firstly, the standard headset cable length (5 m) was not long enough for the purposes of the training application. Myriad Global developers had to purchase several alternative cables, carry out testing, and extend the cabling to 10 m. Given the portability requirement of the VR set up, for use in various locations, the tracking base stations were mounted on to tripods and not fixed to a wall or ceiling and so several tracking considerations arose. If the base stations were moved out of position, accidently or otherwise, the system would need to be re-calibrated and training would have to re-start from the beginning. Another consideration was related to tracking being lost if an individual stepped into the training space and blocked a base station. When this occurred, on occasions, this caused the virtual camera (i.e. the users view) to shake and cause discomfort (e.g. a feeling of the user falling or being dragged to the side with hand controllers 'floating' away from the user's hands). Tracking issues were also experienced where locations were used with excessive reflective surfaces such as glass. Recruits had to ensure training took place in an appropriate space away from windows or glossy furniture.

Interestingly, a unique consideration arose when MG developers discovered one USAF recruit with an eye condition—astigmatism. Solutions, subsequently, were being considered including the use of customised lenses (Tseng et al., 2018).

4.3 Using the Training Application

Positive feedback was received in regard to the training application. In particular, several references were made in relation to the level of VR realism and attention to detail with the inclusion of small markings and tiny labels. Capt. Matthew Perkins, first WXG Science Officer for USAF, provided the following statement:

> The realism of the VR was incredible. I could make out tiny labels and serial numbers on equipment and aircraft even flew overhead during the simulation. Virtual reality brings unprecedented realism to our training ability when the physical equipment is unavailable. Our deployed Airmen will have greater familiarity with these tools than ever before.

A number of considerations did arise as a result of hardware-related limitations that impacted the training application experience. One particular point was raised on the HTC VIVE headset Field of View (FOV) of 110 degrees and, on occasions, this became a limiting factor particularly when working in a large VR space. Occasionally, sourcing tools, or parts, in the large VR setting resulted in users having to carry out more excessive head movements (i.e. more than the required equivalent movement in real-life) to source the items. Where the user was unable to source an object, possibly due to objects being 'lost' or misplaced, there was no functionality to reset the training application from a previous step or time-point. Training would need to be stopped and re-started from the beginning. Finally, the application did not have the feature to save a session for later resumption—training had to be completed in one single take.

Further information on this work can be found on https://www.af.mil/News/Article-Display/Article/1822830/1st-wxgs-virtual-training-brings-real-benefits/ [Apr. 24, 2019]).

5 Future Outlook and Road Map

The USAF's own assessment of this project was that it greatly enhanced training capability around the TMQ-53 and delivered a significant return on investment within the first 6 months of deployment. More broadly it has proven the use of digital twins and immersive VR, in particular, for 'muscle memory' training, analysis and fault finding. The first Weather Group are looking to further enhancements of the solution in the future as the TMQ-53 evolves.

More broadly, MG has been expanding the digital twin concept to address the challenges of training around aviation. Training of engineers in military aviation can be time-consuming and expensive. For example, training on General Electric's F110 turbofan jet engine, used on F-14, F-16 and F-15E platforms, has often been restricted by the availability of engine units to train on. It requires 'hands-on' experience with expensive equipment which can be easily damaged by inexperienced trainees rendering the platforms non-operational.

In common with the TMQ-53 Project, MG are developing a VR application and accompanying blended e-learning solution. This consists of a photorealistic, digital twin of the F110, enabling trainees to learn at no risk to themselves or the equipment. This combined immersive training tool, and e-learning package, is designed to be used alongside the physical training curriculum whilst also supporting conventional classroom and computer-based learning. This will create a safe, standalone or integrated training platform that gives highly realistic visual preparation for engineers, and technicians, preparing them for work on complex engineering procedures on real aircraft.

There are a number of benefits of this approach over more traditional methods, including:

- Improved productivity offering training from a broader range of locations including remote sites.
- Reduction of training time.
- Improved information retention.
- Improved return on investment as a result.

In the energy sector MG are working with complex asset owners (Petrochemical, LNG plants and refineries) to develop VR digital twins again for training, work planning and familiarisation. These are designed to incorporate key elements including reality capture, 3D models, persistent data, building information modelling, and dynamic/real-time data.

Myriad Global are also working on other applications for US and UK Defence organisations, as well as the Energy, Medical and Technology sectors focusing on the deployment of VR for familiarisation, procedures and maintenance. The outlook for this application of immersive technology is positive given the challenges that many organisations face. These include finite resources around the availability of physical assets, the need to train and work remotely, and the challenge of reducing training time, improving information retention and reducing cost (e.g. Gallerati et al., 2017).

6 Conclusions

Overall, the project, and use of the VR training solution was regarded as a success by the client and users. The client's requirements were achieved from a delivery point of view and there was a significant return on investment achieved.

Specifically, the goals of reducing training and equipment maintenance costs, whilst increasing efficiency of training and improving retention of related USAF personnel, were achieved. The USAF has quantified the cost saving at approximately two million USD in the first six months of deployment.

The approach of creating a virtual replica (or digital twin) for training purposes is growing as an effective alternative to more traditional methods. With the challenges of traditional methods only likely to continue (or become even more constrained), and immersive capabilities continuing to improve consistently, organisations across a range of sectors are reviewing their current solutions and looking towards innovation.

The ability for trainees to learn the required skills and gain the required experience, combined with the benefits of reduced deployment costs, reduced equipment wear, flexible deployment options and the reduced risk to human safety, will lead to a significant increase in the number of organisations experimenting with, and adopting, virtual training solutions into the future.

References

Bell, B., Kelsey, E., & Nye, B. (2020, July). Adapting instruction by measuring engagement with machine learning in virtual reality training. In International conference on human-computer interaction (pp. 271–282). Springer, .

Fuller, A., Fan, Z., Day, C., & Barlow, C. (2020). Digital Twin: Enabling technologies, challenges and open research. *IEEE Access, 8*, 108952–108971.
Gallerati, P., Bagnato, S., Casciaro, D., Conte, A., & Maisano, M. (2017, May). Use of serious gaming and virtual reality applications improves students' learning retention and reduces safety risks and costs associated with training activities. In *Offshore mediterranean conference and exhibition*. Offshore Mediterranean Conference.
Roles, A. F. W. (2017). By order of the commander 305th Air Mobility Wing (AMC). *Weather*.
Tseng, S. P. J., Richards, E. M., & Shen, S. (2018). *U.S. Patent No. 9,995,940*. Washington, DC: U.S. Patent and Trademark Office.

A Preeminent Digital Production Studio Collaborates with One of America's Most Innovative Universities to Explore the Use of Virtual Reality for Learning

Adam Blumenthal and Faby Gagné

Abstract

Southern New Hampshire University is a large, innovative institution of higher education in the United States. In addition to a physical campus, the university also serves over 120,000 students online, offering Bachelor's Degrees, Master's Degrees, Associate Degrees and Certificate Programs. Due to the large population of remote learners the university has developed sophisticated digital platforms for teaching and learning. SNHU also monitors trends in education and technology. With an interest in virtual reality's (VR) potential to impact teaching and learning, the Innovation Center at SNHU partnered with Nucco Brain at Unit9, a leading digital innovation and production studio, to design and build VR-based active learning experiences for a neuroscience course. The two teams utilized a collaboration methodology with faculty subject matter experts and learning and instructional specialists working closely with creative technologists.

Keywords

Virtual reality · Immersive learning · Higher education · Neuroscience

A. Blumenthal (✉)
Nucco, a Unit9 Company, London, UK

Brown University, Providence, Rhode Island, United States
e-mail: adam.blumenthal@nuccobrain.com

F. Gagné
SNHU Labs, Innovation Center, Southern New Hampshire University, Manchester, USA
e-mail: f.gagne@snhu.edu

T. Jung, J. Dalton (eds.), *XR Case Studies*, Management for Professionals,
https://doi.org/10.1007/978-3-030-72781-9_13

1 About the Collaborators

In the Fall of 2018, Southern New Hampshire University (SNHU), one of America's most innovative[1] and largest universities, published its *2018–2023 Strategic Plan.*[2] Embedded in this plan is a commitment to a series of initiatives that will help them create the capacity and foundation on which to build a *2030 learning ecosystem*, with new forms of interactive learning, integrated student assessment, and personalization. SNHU worked with *The Institute For The Future* in Palo Alto, California to envision what the world, society, and education might be like in 2030 which helped frame and operationalize its research and development experimental program.[3]

But it is not just recently that SNHU has considered the future and taken action to serve new generations of learners. The university, with a 300-acre campus in Manchester, New Hampshire, launched one of the first online colleges in 1995, in the earliest days of the world wide web. The evolution of that first online college is a global program that now offers bachelor's and master's degrees, associate degrees, and career-oriented certificates to approximately 130,000 students, making SNHU one of the twenty largest institutions of higher education in the United States.

As a long-time digital disruptor, SNHU understands the value of being ahead of the curve, and invests in that. Their Innovation Center and research and development (R&D) unit, SNHU Labs, is experimenting with promising technologies with the goal of specifying the requirements and roadmap, and developing the platform that will serve the learner of 2030. One of the promising technologies SNHU is exploring is virtual reality for immersive learning.

SNHU's 2018–2023 Strategic Plan also calls out the need for the university to form partnerships and collaborations with service providers with high levels of expertise to support their innovative work. This call led to a collaboration between SNHU Labs and Nucco Brain, a digital innovation studio that's part of the Unit9 Group. Founded in 1996, Unit9 is a family of studios with specialties in digital innovation, strategy, design, technology development and R&D, film and animation, and sound production. Within the group, Nucco Brain specializes in designing and building innovative digital learning experiences. With a headquarters in London, Unit9 and Nucco Brain operate from three offices in Europe and three in the United States. In 2019 AdAge named Unit9 "Production Company of the Year."[4] In 2020 Campaign Magazine awarded Unit9 "Tech Company of the Year."[5] The group has developed digital programs with dozens of the world's best known companies,

[1]https://www.usnews.com/best-colleges/rankings/regional-universities/innovative

[2]http://bit.ly/SNHUStrategicPlan

[3]*Innovation Center (2020). SNHU's Futures Strategy.* Southern New Hampshire University.

[4]adage.com/article/special-report-agency-list/ad-age-2019-production-company-year-unit9/2163426

[5]https://www.campaignlive.co.uk/article/campaign-tech-awards-2020-tech-company-year/1686764

organizations, and institutions, earning hundreds of awards, including 34 Cannes Lions, the most prestigious award in the digital creative arts. Nucco Brain and Unit9 have produced immersive learning experiences with organizations including Anglo American, Arizona State University, EDF Energy, John F. Kennedy Presidential Library and Museum, Resuscitation Council UK, and Samsung, among others.

2 Project Summary

With both their campus and online student bodies in mind SNHU began exploring the transformative potential of teaching and learning with virtual reality. Before SNHU selected Nucco Brain/Unit9 as its design and development partner, SNHU Labs authored a *Charter for a pilot project for immersive learning experiences.* The pilot project had three research goals:

1. To investigate the potential of immersive learning experiences enriching student experiences;
2. To investigate whether supplementing typical teaching and learning with VR-based approaches improves learning outcomes; and
3. To understand the process of producing an immersive learning experience in collaboration with an external partner and the feasibility of implementing VR in the classroom.

The *Charter* helped sharpen the short-term goals and long-term vision for the project. Once the design and development phases were underway it helped the combined teams maintain focus on the project's research objectives, and limited the possibility of introducing additional scope.

SNHU Labs was interested in connecting the pilot project to an active course. They met with deans and faculty from the university to introduce the pilot project. They sought subject matter that is especially complicated or challenging to teach or learn using textbooks and lectures, and they wanted a professor who could engage with the combined design team as a subject matter expert and, in general, to be a champion of the pilot effort. Lastly, SNHU Labs wanted to be able to work with students who were actively engaged in the course material so the app that would be the outcome of the pilot project could be tested within the context of a typical semester. SNHU Labs chose to work with Dr. Peter Frost, a professor of psychology in the SNHU School of Arts & Sciences. Professor Frost's course, *Biopsychology*, was selected. In this course students explore the relationship between the brain and the nervous system and their connections to behavior.

Thus the collaboration set out to produce *SynapseVR*. SNHU Labs maintained a collaborative mindset throughout the project, recognizing that the production of innovative learning experiences require expertise from the academic side (e.g., faculty, student, and executive support, instructional and assessment designers, and subject matter expertise) and the creative-technology side (e.g., narrative design and copywriting, visual and sound design and production, iterative development, and production management).

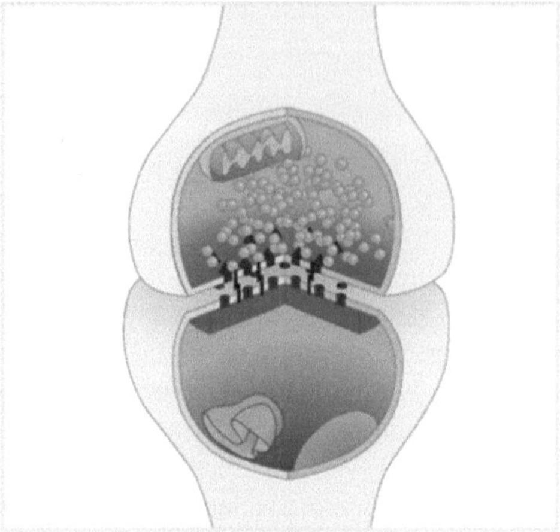

Image of a synapse, typical of a textbook illustration. https://commons.wikimedia.org/wiki/File:Neuron_synapse.png

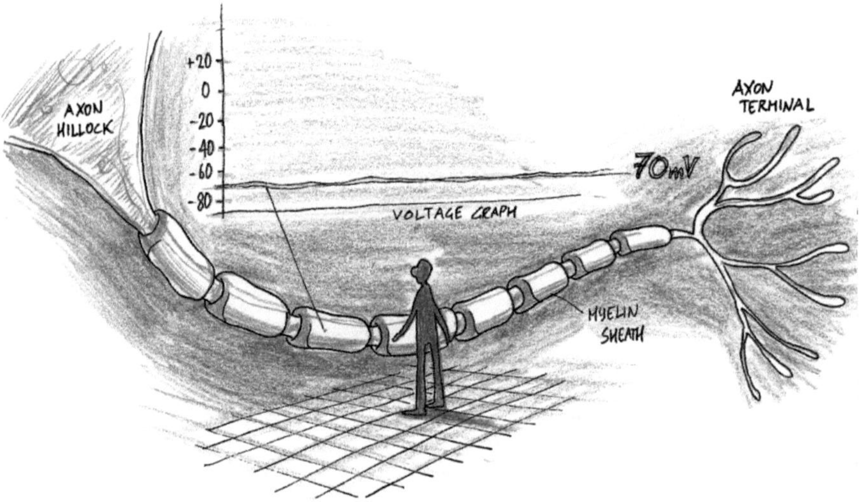

Storyboard illustration for *SynapseVR*

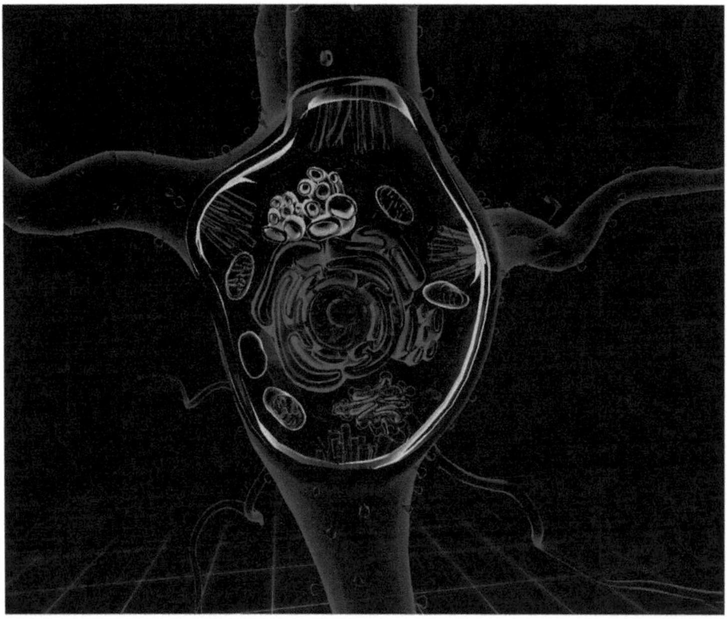

Interactive 3D Synapse from *SynapseVR*

3 Project Details

With the course and professor selected by the university, SNHU looked to Nucco Brain/Unit9 for its recommendation on the VR hardware to be used. The recommendation was the Oculus Quest, a wireless VR system (a connected computer is not required), affording users "six-degrees of freedom," that is, the option to explore 3D space by walking forward and backward, looking high and low, and gazing around in 360-degrees. Furthermore, with its two hand controllers the Quest allows users to reach out and interact with digital objects, a feature that supported the goal of offering students new ways of learning complex material by allowing them, for example, to play a role in controlling systems in order to discover how they work, where the alternative teaching forms (reading and lectures) are quite challenging for students to gain the understanding. The Quest is a relatively low-cost, high-performance system that's easy to use and maintain, minimizing the requirement for extensive hours of technical support when deploying for user testing and classroom use.

While the *Charter* defined the research objectives of the pilot, Seth Corrigan, SNHU's Senior Director of Assessment and Game Based Learning, worked with colleagues and Professor Frost to establish the learning objectives for *SynapseVR*. The team designed the experience using a developmental approach and an existing cognitive model to inform learning design. The cognitive model laid out a progression of increased understanding students should gain from using the experience. Elements of game based assessment and "productive struggle" were incorporated into the design. *SynapseVR* was instrumented with telemetrics (data analytics) to support real-time formative assessment and to measure feedback from student performance. Paper-based prototypes were used with student testers to convey delivery of several aspects of the desired learning experiences in support of early user-testing that guided digital design decisions. Lastly, the prerequisite knowledge, skills and abilities for the student users was described and Nucco Brain/Unit9 was provided with student persona profiles to understand the demographic, psychographic, and technographic characteristics of the student users prior to beginning the design.

With the learning design established by SNHU, Nucco Brain/Unit9 led the process of designing the look and feel and the user experience (UX) of the app. Prior to diving in to design, the teams jointly reviewed the course content and target learning objectives in order to identify the specific affordances of VR that are best suited to foster student achievement of the learning objectives. Then Nucco Brain/Unit9 began to establish the narrative experience for the app. Visual styles were explored, and a design direction decided upon by the collaborators, then copywriting began - all in close collaboration between the two teams and the professor. An iterative development process was utilized in which several builds were tested with students throughout the production. During user testing, managed by SNHU with students on their New Hampshire campus, testing facilitators captured two forms of video from each testing session:

1) an observational video that recorded the student tester in the room, interacting with the app;

and 2) a video capturing the student's first-person perspective from inside the app where stakeholders could watch what the student was seeing and doing.

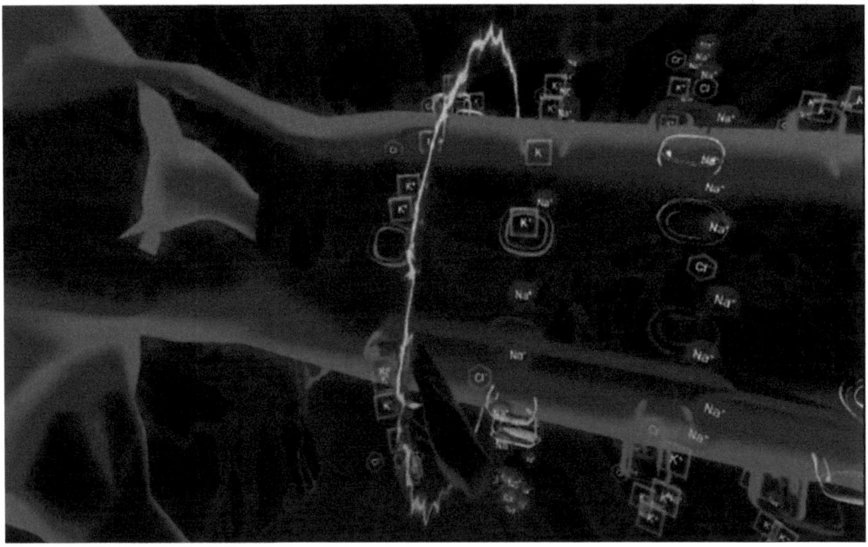

These two types of video recordings, as well as student feedback and stakeholder observations, gave the team the opportunity to adjust characteristics of the design or otherwise to validate the design choices and move forward with a degree of confidence.

4 Results

Three research questions were examined by the SNHU Labs team:

1. Does supplementing typical teaching and learning tools (i.e., textbook reading and lectures) with the use of *SynapseVR* improve learning outcomes?
2. Does VR enrich student experiences as measured by student-reported levels of interest, motivation, and engagement with *SynapseVR*;
3. What conditions are needed to feasibly implement VR in the classroom?

To understand the efficacy of using VR as a supplement to support improved learning outcomes, a rigorous randomized controlled trial (RCT) was designed. A first group of students (the control group) will be taught the Biopsychology material covered in *SynapseVR* using the traditional methods (textbook and lectures), and a second group will have access to these traditional methods plus the *SynapseVR* supplemental experience. The RCT will allow SNHU testers to understand the degree to which students in the second group performed better or not than students in the control group. At the time of this writing the COVID-19 pandemic has caused the postponement of the RCT, previously scheduled for the spring of 2020.

However, several other feedback and testing mechanisms to address the three main research components were designed and deployed prior to the restrictions of COVID-19. In February of 2020 SNHU Labs worked with two of Professor Frost's classes to examine classroom implementation. Twenty-one students were put into groups of two or three with "explorers" (those in the VR experience) and "guides" (those students supporting the explorers). The students switched places so all students had the opportunity to be both explorers and guides. Repeated measures testing suggest that students' content knowledge was higher following the VR learning experiences than when it was measured at baseline. In addition, results were compared between one of Professor Frost's 2019 classes (students who did not have access to VR) to results from his 2020 classes of students who used *SynapseVR*. There is evidence from this initial study that VR increased learning - however the results were mixed and only evidenced in the more conceptual test questions. Results from the RCT are still needed to fully understand the impact of *SynapseVR* on learning.

As for the potential of VR to enrich student experiences as measured by student-reported levels of interest, motivation, and engagement with *SynapseVR,* surveys were used to pose fifteen questions to the testers. With seventeen participants, students reported high levels of interest, motivation, and engagement:

1. Nearly 90% of the students agreed with the statement, "I enjoyed learning this way;"
2. Approximately 80% of the students agreed that "I would like to learn this way in the future."
3. 80% of the students felt motivated to "learn more about this subject."
4. More than 90% of the students "felt that the lesson was engaging" and "found the lesson to be useful."
5. Approximately 80% of the students agreed with the statement, "I felt motivated to understand the material."

In terms of their emotional response to *SynapseVR,* students reported high positive and.

low negative affective states:

1. 80% of the students "felt happy during the lesson."
2. About 90% of the students "felt excited during the lesson."
3. More than 90% of the students reported that they did not "feel bored during the lesson."

In summary, Cat Flynn, SNHU Innovation Center's Deputy Director of Learning Science, concluded, "the students overwhelmingly loved the experiences with *SynapseVR.*"

To understand the feasibility of implementing VR in the classroom, SNHU Labs again worked closely with Professor Frost to develop resources for instructors, explorers and guides, tested these methods with "implementation rehearsals," and then observed several aspects of implementation during the classroom testing sessions. Below are some of the key findings:

- Most students were relatively inexperienced using virtual reality, and some required additional on-boarding support before they were ready to engage with the learning experience. The team concluded that more on-boarding time is needed so that when the immersive learning experience begins, students have no barriers to participation.
- SNHU Labs produced a "Guidebook," which was considered a valuable resource by all participants. It offered troubleshooting tips, advice for getting acquainted with interacting in the virtual space, and recommendations for health and safety.
- During student testing technical support personnel were on-hand to address any technical issues that arose, so students could return to the experience expeditiously.
- The integration of well-considered telemetry provided a wealth of data to analyze some quantitative measures related to student behaviours and engagement.

5 Future Activities

At the earliest opportunity, SNHU will run the RCT for a more rigorous study of the learning efficacy of VR. The team is also preparing a *VR Playbook* summarizing research findings and offering practical recommendations for implementing VR at SNHU. Both parties, SNHU and Nucco Brain/Unit9, agree that an effective model for collaboration between academia and production studio was established, while citing the following observations:

- It's best not to rush early phases of the design framework such as identifying clear descriptions of the content to be treated or identifying the desired learning outcomes. Since the content topics are potentially complex, spending more time in this phase of the work reduces the likelihood of disruptive changes later in the project;
- It was important for Nucco Brain/Unit9 to understand SNHU was seeking a collaborative relationship in which both groups would be involved in the work. Because of this collaborative setup, Nucco Brain/Unit9 proactively helped the SNHU research team solve for internal validity issues with respect to the design of the efficacy study;
- To make sure SNHU team members were not providing conflicting feedback to the creative-technology team the "one voice" approach was utilized for all feedback in which all feedback was funneled through a single person to be sure the feedback messages were consistent;
- Harness the voice of the subject matter expert: because the Innovation Center team and Nucco Brain/Unit9 are not expected to be experts in biopsychology (or any other number of disciplines), the faculty SME was crucial to ensuring that

the experience was scientifically accurate and connected with existing course materials;

- It would have been helpful to the SNHU team to have a "menu" of the types of interactions that are possible in VR in advance of the learning design work, as well as a glossary of relevant technical terms to establish that all collaborators are on the same baseline;
- Given this was the Innovation Centers's first design and development project in virtual reality, there were no established processes to help streamline some of the work and decision points at various stages of the project. Working with an experienced vendor with prior successes helped accelerate SNHU's learnings.

The two teams have already begun a second collaboration of a different sort: a software experience that uses speech and natural language processing to analyze and provide feedback to young adults who are introducing themselves in the context of a mock job interview.

Acknowledgments **SNHU Innovation Center & SNHU Labs**

- Executive Sponsor: Wil Zemp, SNHU Executive VP and Chief Strategy and Innovation Officer
- Faby Gagne, Executive Director, SNHU Labs
- Seth Corrigan, Senior Director of Assessment and Game Based Learning
- Cat Flynn, Deputy Director of Learning Science
- Cait Glennen, Project Manager
- Alan Mendonca, Graduate Assistant
- Jesse Damiani, Deputy Director of Emerging Techology

SNHU School of Arts & Sciences

- Professor Peter Frost
- Dean Steven Johnson

Nucco Brain/Unit9

- Stefano Marrone, Managing Director
- Adam Blumenthal, Executive Producer
- Ryan Lovelock, Creative Director
- Łukasz Czechowicz, Developer
- Amanda Ramos, Producer
- Jenny King, Producer
- Andy Thomas, Copywriter
- Johnny Leal, Animation Lead
- Eva-Gina Berkel, Audio Production
- Jan-Niklas Kuhn, Audio Production
- Raggie Drosshaug, Production support

Reference

Biopsychology textbook. (n.d.) *An Introduction to Brain and Behavior*, Fifth Edition, by Bryan Kolb, Ian Q. Whishaw, and G. Campbell Teskey.

Virtual Reality's Role in Improving Student Knowledge and Engagement in the Materials Testing Laboratory

Rajesh Jha, Douglas Choi, Rebecca Traboulsi, and Arun Srinivasa

Abstract

The aim of this work was to evaluate the efficacy of using virtual reality (VR) to develop training modules for students in mechanical engineering laboratory classes. Specifically, we investigated the feasibility of applying VR in material testing labs to increase engagement and learning outcomes and also in fostering more effective use of expensive lab equipment. To validate the VR software and test our hypotheses, we conducted pilot testing at Texas A&M University. A two-group, randomized control trial was conducted with 118 students. Results of the pilot test suggest an overall positive outcome for students participating in the VR virtual lab first (experimental condition) compared to those who performed the physical lab first (control condition). Students in experimental condition scored significantly higher on both multiple-choice and open-ended subject matter questions. Subgroup analysis indicated that females in the experimental group gained twice as much as males.

Keywords

Virtual reality · VR · Training · Self-efficacy · Engagement · Mechanical engineering

R. Jha (✉) · D. Choi · R. Traboulsi
SimInsights Inc, Lake Forest, USA
e-mail: rkjha1@siminsights.com; douglchoi@siminsights.com

A. Srinivasa
Texas A&M University, College Station, USA
e-mail: asrinivasa@tamu.edu

T. Jung, J. Dalton (eds.), *XR Case Studies*, Management for Professionals,
https://doi.org/10.1007/978-3-030-72781-9_14

1 Company Overview

This case study is the result of the efforts to develop and pilot a VR solution conducted by three organizations: SimInsights, Texas A&M University (TAMU), and Instron.

1.1 SimInsights

SimInsights Inc. is a California based XR software company transforming training by leveraging virtual and augmented reality (VR/AR), simulation, artificial intelligence and machine learning (AI/ML) across employee, customer and product life cycles. Our mission is to enable all subject matter experts to author digital twins for training by extracting domain knowledge from existing 3D, text, images and videos without programming/ML/AI skills.

1.2 Texas A&M University (TAMU)

Texas A&M opened its doors in 1876 as the state's first public institution of higher learning. Today, TAMU stands as a research-intensive flagship university dedicated to sending "Aggie" (name for students and alumni) leaders out into the world prepared to take on the challenges of tomorrow.

1.3 Instron

Founded in 1946, Instron® is a recognized worldwide market leader in the materials testing industry. Instron's goal is to provide customers with the best ownership experience by delivering the highest quality products, expert support and world-class service.

2 Project Summary

First, we developed new software modules and data representations for virtual objects, scenes (simulations), tasks, instructions and student performance, and applied them to build a novel software product called HyperMock to show the feasibility of enabling non-programmers to create high quality VR based training simulations. We used HyperMock to create most of the tensile test simulation shown in Fig. 1 except the specimen deformation and stresses which were computed using Abaqus finite element analysis (FEA) simulation software from Dassault Systemes, and imported into HyperMock. We pilot tested the VR simulation as part of a course at TAMU. The total number of students with complete data analyzed for this study was 118. A two-group, randomized control group experiment was conducted.

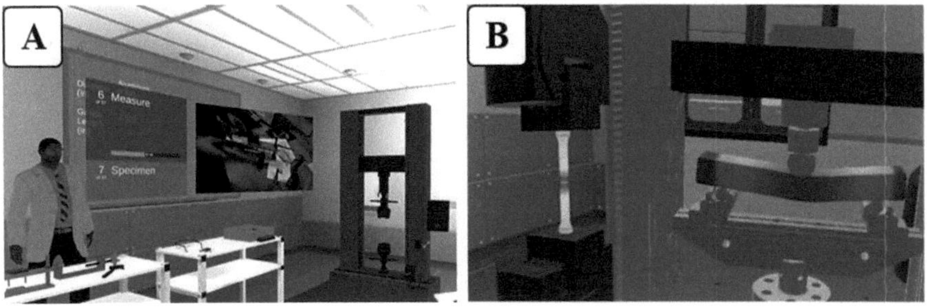

Fig. 1 (**a**) User's view of the VR simulation showing pedagogical agent, text and video instructions, and Instron machine. (**b**) Zoomed in view of VR simulation showing tensile and three-point bending tests. Contours show Abaqus nonlinear FEA results

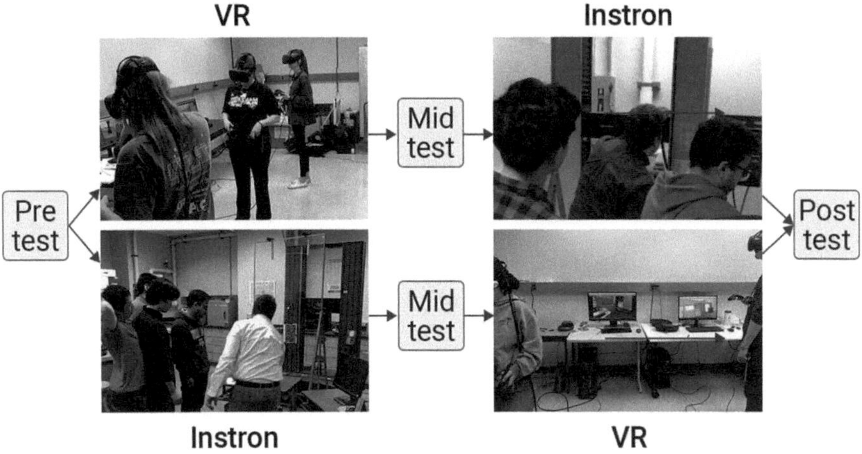

Fig. 2 Study design used for Phase 1 pilot. Top path is the experimental group. Bottom pathis the control group

Figure 2 illustrates the study design. 57 participants were assigned to the control condition (Instron machine is used first, followed by VR) and 61 to the experimental condition (VR first, then Instron).

The key outcome metrics were engagement and learning outcomes (procedural and conceptual, measured using multiple choice and free response questions).

3 Project Details

3.1 The Situation

We selected the tension test due to the high equipment cost and inherent safety risks involved in high tension material testing. Student errors can cause damage to expensive machines resulting in tens of thousands of dollars in repair costs as well as significant downtime and lab scheduling disruptions. TAMU faculty were looking for ways to improve lab training while reducing the overall cost and safety risks. Specifically, the requirements were:

- Replicate the experiential learning of physical labs while eliminating the safety risks, equipment availability bottlenecks and costly damage risks.
- Increase student engagement
- Improve learning outcomes

3.2 The Solution

In 2017, SimInsights was awarded a small business innovation research (SBIR) award by the National Science Foundation (NSF) to conduct R&D on the feasibility of VR/AR enabled personalised training in manufacturing. SimInsights teamed up with TAMU and Instron and began work on the following fronts:

- **An immersive approach to virtual labs—**with an emphasis on providing engagement and experiential learning that rivals or exceeds physical labs, a vision of investing in the latest technologies, and the aim of transferring the knowledge and skills to projects in school and beyond.
- **Randomized controlled trial (RCT) based evaluation—**which aims to rigorously assess the impact of interventions in human subject studies where many confounding factors may be present.
- **Innovative authoring software for immersive training**—called HyperMock to enable faculty and other subject matter experts to create their own immersive content.

3.3 Details of the SimInsights VR System

SimInsights VR system consists of a VR headset and the HyperMock application for authoring, publishing, experiencing and evaluating training simulations. HyperMock provides the following key capabilities:

- Ingest existing 3D, PDF, image or video-based content and package it into a reusable virtual object (VO) representation

- Organize the VOs in an online repository so they can be reused across training scenarios
- An application with a graphical user interface that non-programmer subject matter experts (SMEs), such as university professors, can use to *author* training simulations that include step by step instructions
- Publish the simulations to students across web, VR and AR platforms
- Enable the students to access and experience the simulations
- Telemetry to collect detailed interaction data for analysis of engagement and learning outcomes and to support personalization in future
- Support voice user interactions (VUI) so that students can speak with pedagogical agents (PA) within the simulations

3.4 Why TAMU Was Attracted by the SimInsights VR System

TAMU had been using video-based instruction as part of pre-lab activity but that approach had not produced the optimal level of engagement and learning. An immersive VR based approach offered the following potential benefits:

- **Immersive experience**
  TAMU faculty wanted to retain the maximum experiential equivalence between the physical and virtual lab sessions, but without the safety risks and equipment bottlenecks. Specifically, the instructors wanted to enable students to:
  - Follow the lab procedures step by step and remember how to correctly use various equipment as well as avoid damage to equipment
  - Become aware of potential safety risks
  - Learn concepts related to the experiment
- **Incorporate the physics of material behavior**
  Physical labs do not allow visualization of the internal changes occurring within test specimens which limits conceptual learning and deeper understanding. Faculty aimed to circumvent this limitation by using finite element analysis (FEA) simulation for in-situ visualization of what is happening inside the test specimen during testing, so that the VR simulation could deliver new insights not possible with the physical test.
- **Conversational AI based pedagogical agents**
  Universities everywhere are looking for ways to expand access to quality education and TAMU is no exception. If AI can enable a "virtual teaching assistant" to provide guidance to students in lab activities, that would go a long way toward expanding access to quality training. To this end, we added a virtual person (VP) with realistic and natural preset animations such as walking, looking at, talking, turning, etc. to play the role of a pedagogical agent (PA) (lab assistant) who could assist students by responding to spoken questions by voice as well as relevant animations. In response to the question "Where is the caliper?", the PA can highlight the caliper using an arrow.

3.5 How SimInsights VR System Was Integrated into the Curriculum

We set up three VR systems in a separate room in the engineering building (see Fig. 2 above). Instructors from different classes are able to reserve the room for their students. Students can access the room at their own convenience, log in and experience the VR simulations as many times as needed. The data can be collected and shared with the relevant instructors. Beside the initial materials course, additional faculty have used the set up for their courses including Industrial Engineering, Chemical Engineering and Engineering Design.

3.6 Evaluating the Impact

The VR program enjoyed considerable success as measured by student engagement and learning outcomes. Specifically, analysis of pilot test data (survey measures, action and sensor data) from 118 students revealed the following highlights:

- **23% greater learning gains on multiple choice questions (MCQ)**—Students who participated in the VR activity first scored 23% higher on MCQ subject matter scores, including 5% higher procedural subscale and 49% higher conceptual subscales scores.
- **106% greater learning gains on free response questions (FR)—**Students were asked, both before and after the activity, to list the steps in the experimental protocol and were awarded one point for each correct step. Analysis of pre-post gain indicated significantly higher gains for participants in the VR activity compared to the Instron activity.
- **15% greater engagement—**Additionally, students reported significantly higher engagement after the VR activity than after the physical lab.
- **100% greater learning gains for female participants**—Subgroup analysis of pre-post learning gains indicated that females in the experimental group gained twice as much over males.

About ten faculty from different courses and departments were given demonstrations of the system. Many of them participated in a one-day workshop to discuss potential applications in their various courses and subsequently engaged with SimInsights to adopt VR in their courses. See sect. 3.5 for the courses and sect. 4 for their feedback.

4 Feedback from End Users

The SimInsights VR system is considered a clear success for TAMU. The user survey included a section for students' comments. Some sample comments are included below:

Question	Sample responses
Overall comments or feedback	*"This system was life-like and I loved how I was able to basically do the real life application using this system. This would be amazing in the future!"* *"The experience is extremely enjoyable and positive."* *"Experienced some difficulty because I wear glasses but overall a great experience."* *"Very helpful in learning the process. They should have one of these things around at all times"* *"Very very positive experience. The only issue was that the head set felt a little bit weighty."* *"Virtual reality simulation was awesome. When viewing the specimen under stress it felt like something out of iron man"*

Following is a sampling of the feedback from faculty members and industry partners:

> Our faculty have been developing VR and AR content for use in our mechanical engineering courses at Michigan Tech and it has been a pleasure to work with the SimInsights team and their software platform. VR/AR technology has tremendous potential for the future of education and research.

William Predebon, Professor and J. S. Endowed Chair of Mechanical Engineering-Engineering Mechanics, Michigan Technological University.

> I have been using the HyperMock CNC machining simulation in my courses in Industrial engineering at Texas A&M for the last couple of years. It is both an engaging experience for students as well as offers substantial educational value for the targeted content. Furthermore, the VR simulation is a nice tool to get students thinking about innovative applications of this exciting technology in other domains.

Dinakar S., Assistant Professor, Industrial Engineering, Texas A&M University

> I have been developing a chemical plant control VR simulation for use in my courses in Chemical Engineering at Texas A&M and it has been a pleasure to work with the SimInsights team and their software platform. VR has tremendous potential for the future of education.

Joseph Kwon, Assistant Professor, Chemical Engineering, Texas A&M University

> It was my pleasure to support this project by SimInsights from the Instron side. We provided CAD models and technical information which enabled SimInsights to create a realistic virtual simulation of the Instron machines. Later the team visited our headquarters in Massachusetts to demonstrate their solutions. This technology has a promising future for training and sales purposes.

Britt Smallwood, Business Unit Manager, High force applications, Instron.

5 Future Outlook/Roadmap

Looking ahead, our aim is to continue development of VR content in partnership with faculty to progressively add more scenarios and respond to evolving needs and priorities (such as adding support for web and desktop to support social distancing and online access).

The long-term aim is to produce a complete collection of gamified training modules requested by our academic customers, potentially moving beyond student training into employee training and even recruitment. VR will eventually be deeply integrated into the way the university trains its students and employees.

6 Conclusion

The use of SimInsights VR system has made a significant contribution to the courses at TAMU and provided a proof of concept to emulate in other universities. In fact, upon learning of this study, two universities—Michigan Technological University and University of California, Irvine—began working with SimInsights to implement VR in their curriculum.

From SimInsights' perspective there were four key factors that contributed to the success of the project, as described below.

6.1 Success Factor #1: Content First

In education, as in other industries, content is king. To be considered seriously for inclusion in curriculum, the VR simulation needed to match or exceed the physical lab experience in *fidelity* and engagement. The SimInsights team delivered on this requirement by matching the realism of the interactions and exceeding the richness of the learning by using nonlinear finite element analysis (FEA) simulation to visualize the physics of material testing. All the assessments used in the study were developed with *validity* as the key objective. This "content first" approach was necessary to win the trust and approval of faculty and get them excited about adopting VR.

6.2 Success Factor #2: Data Driven

Starting with the early usability studies conducted with TAMU students, SimInsights followed data to drive prioritization and design of the software features and functionalities. Three usability studies helped to progressively refine the prototype and prepare for the final pilot test. Considering that VR is an entirely new medium, the design space of experiences is vast. Data can be an effective guide in the search for the optimal experience.

6.3 Success Factor #3: Conversational AI

With personalized training as our ultimate goal, conversational AI has always been a key element of our VR experience design strategy. Whereas the faculty were unsure about the utility of this interaction element, students loved it from the very beginning, providing valuable feedback to drive refinements, and challenging the team to support increasingly complex and meaningful queries. Adding conversational AI enabled us to provide a "wow" factor that went way beyond immersive visualization. It also opened us to the rich possibilities of multimodal AI that transfer very well from VR to AR.

6.4 Success Factor #4: Partnering with the Experts

Perhaps the most important success factor was that we recognized the importance of partnering with experts early on, scheduling weekly meetings with key faculty and even teaching assistants. This enabled our team to produce content that addressed key learning and assessment objectives, without wasting precious resources on "gimmicky" nice to have features. Experts are truly the cornerstone of our product strategy. To enable them to capture their knowledge and skills using our authoring tools is the guiding principle behind our product strategy.

Bank of Ireland Tests Meetings in Virtual Reality

Abraham G. Campbell, Athanasios Staikopoulos, Thomas Holz, Mike Harlick, and Jonny Cosgrove

Abstract

This paper summaries a case study conducted by Meetingroom and Bank of Ireland in exploring the use of virtual reality to conduct meetings online. The case study was to compare virtual reality meetings to traditional video teleconferencing software. The results point to the improved feelings of presence, closeness, and arousal for virtual reality environments, but also demonstrate a difference in the experience for female and male participants. In particular, the use of avatars instead of real-life video images of the participants was preferred by female participants which point to a possible de-biasing ability for virtual reality environments.

Keywords

VR collaboration · Video conferencing

1 Company Description

MeetingRoom Software Limited operates meetingRoom.io, a platform that enables team communications in one place. The platform provides a range of capabilities, including real-time and contextual group chats using spatial voice calls, file storage and sharing. In the future it will incorporate integration with various enterprise apps, and direct messaging and searching. Meetingroom provides virtual meeting rooms in

A. G. Campbell (✉) · T. Holz · M. Harlick · J. Cosgrove
MeetingRoom, Dublin, Ireland
e-mail: abey@meetingroom.io

A. Staikopoulos
Learnovate Centre, TCD, Dublin, Ireland

T. Jung, J. Dalton (eds.), *XR Case Studies*, Management for Professionals,
https://doi.org/10.1007/978-3-030-72781-9_15

the cloud. Traditionally these virtual reality cloud platforms are used for gaming rather than enterprise environments but ironically the requirements for gaming networks in terms of security and reliability have now acceded that of many existing enterprise platforms. The advantage of using a cloud-based approach to network is primarily scalability and that it allows the support of thousands of users with ease.

2 Project Summary

Today's workforce environments are steadily becoming more distributed across the globe, raising the need for improved ways of facilitating collaborations at a distance (Oprean et al., 2018).

Virtual worlds are expected to make an evolutionary leap to the workplace and reshape business collaboration and communication even more profoundly than videoconferencing did 10 years ago.

Virtual reality applications enable employees to get as much mobility and flexibility as they desire, and autonomy to work and access a real visual representation of their working environment. Virtual meeting rooms can be used to simulate real-life situations and show how people behave, gesture and move around during conversations.

These technologies are powerful tools in the meetings and events space, because they allow the creators to invent a world or recreate an existing space that gives the users a fully immersive experience. In addition, collaborative virtual environments provide users a sharing space supporting different types of collaboration and interaction scenarios.

However, despite the recent VR technology advances, there has been limited research on the usability and effectiveness of Collaborative Virtual Environments (CVEs) (Schroeder et al., 2006). Furthermore, there are even less CVEs which have been applied and evaluated in the context of working and meeting spaces. One of the main reasons was the immaturity of the technology causing many usability problems. Another reason was that CVEs were not available to users for regular use (Schroeder et al., 2006). In addition, most of the CVEs were developed as prototypes and proof of concepts for research purposes. In general, the usability and effectiveness of CVEs is related to the degree in which participants can remotely meet, communicate, collaborate, interact and complete their assigned tasks effectively and from within a virtual working environment like in real life.

The purpose of this experiment was to evaluate the efficiency of representative virtual reality collaborative tools compared to a video conference. The case study is summarised in this paper but if the reader is interested in greater detail on any aspect please consult "Uses of Virtual Reality for Communication in Financial Services: A Case Study on Comparing Different Telepresence Interfaces: Virtual Reality Compared to Video Conferencing" (Campbell et al., 2019).

3 Project Details

The research was conducted using meetingRoom.io meeting software and Skype for Business through an A/B test at the Bank of Ireland in Grand Canal Dock. Facilitated by University College Dublin with 100 participants, the project setup for one of the VR stations is shown in Fig. 1.

Two participants will be meeting using either VR or Skype, then they will be surveyed on their experience. Following that experience, the users will then try to meet again using the alternative method and be surveyed again. The participants will be using a media experience survey along with a questionnaire asking them which they preferred.

The research hypothesis is that the VR experience will make the users feel more in control, be more present with the other user, and more enjoyable than the video link up.

In summary, the case study considered several hypotheses:

- Virtual reality will be a more positive experience.
- Participants will feel more in control of the application in virtual reality
- Participants will feel more immersed in the VR meeting
- Participants will feel closer to each other in the VR meeting
- Participants will find the virtual reality meeting more exciting than the video conference

Fig. 1 Paricpant using HTC VIVE VR setup during the experiment

4 Feedback from End Users

The feedback from end users is outlined in Fig. 2 below. The participants used a media experience survey along with a questionnaire asking them which they preferred. The VR users felt more in control, more present with the other user, and found the experience more enjoyable than the video link up. The results point to the improved feelings in using virtual reality environments of:

- Presence
- Closeness
- Arousal

The study also clearly demonstrated the difference in the experience for female and male participants. In particular, the use of avatars instead of real-life video images of the participants was preferred by female participants which point to a possible de-biasing ability for virtual reality environments. The de-biasing potential is due to the fact that participants as avatars may feeling less judged or be judged in a virtual world than in a traditional video conferencing application.

The results also indicate a need to explore different interfaces designs and discuss lessons from the historical adoption of technology from the financial services community. One very telling post interview exchange demonstrated this clearly to the researcher. For context, all post interviews were done separately, so the participants did not know what the other had said as their participation was anonymous. A female participant reported that she felt that the other avatar was not able to intimidate her as would have been possible in a video call. A separate male participant reported that he disliked the virtual reality world as he felt he was not able to intimidate the other participant which he was able to do using a video conference.

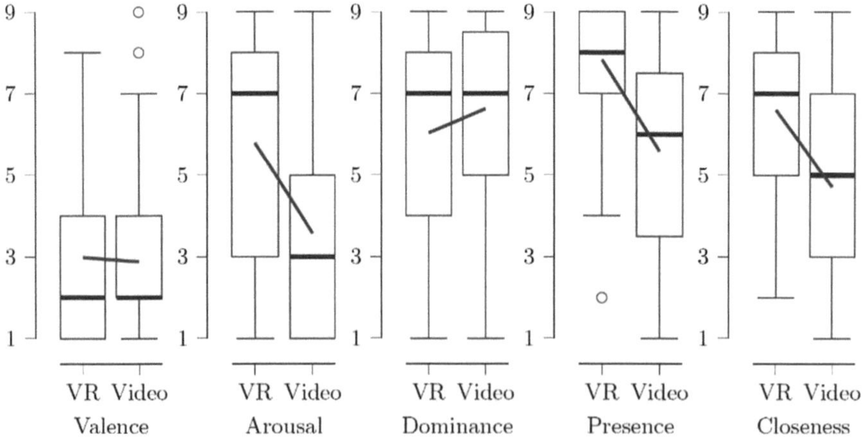

Fig. 2 Results showing the feedback from users in each condition

This is an area that needs to be researched further and it probably due to current societal issues around gender rather than a unique aspect of VR. Also given that the 100 participants were from business-oriented jobs both customers and employees of Bank of Ireland they represent a is quite different cohort to previous VR experiments which normally involved 18–21 college students, who may not have experienced real life working environments.

During our post interviews with each of the participants, many mentioned the fact that on calls with more than 2 people that the conversions would naturally become more difficult as each participant needed to wait for other to stop talking. This waiting for a moment of silence was required to enter into the conversation and the timing of when to talk became even more complex as more people entered the call.

In a virtual world, just like in real life, sound comes from the direction of the participant talking; thus, this waiting was not required. This is not to suggest that people talk over each other but that a participant could discern which speaker was talking along with making out what was said as the audio came from a different spatial direction. Along with more natural body language cues such as simply raising your hand, this above all else points to a new use of virtual reality to conduct teleconference calls in the future. The value provided to the decision-making process with knowledge management in an organisation is exponentially linked with the real world conversation highlighted above.

5 Future Outlook/Roadmap

Overall, the meetings in both VR and VC demonstrated that teleconference technologies have matured to a point that communication remotely when given dedicated non-congested internet connection is completely achievable in the modern world.

Many participants in post interviews stated how video teleconferences normally suffer problems with connectivity. This is normally not due to the creators of the VC software but to the underlying infrastructure. This is an ongoing problem where users do not have the dedicated upload and download speeds required to conduct a call of decent quality.

VR can alleviate some of these issues due to the lower bandwidth requirements of simply sending transforms for the VR avatar location and limbs. This reduction is on average 1/10 the requirements of sending video. Given the potential for the de-biasing and levelling effect of VR, coupled with the ratings from older demographics, this indicates a huge potential in terms of diversity in the HR process and adoption in the financial services industry.

Further case studies are planned with existing customers of Meetingroom to identify in detail where virtual reality collaboration meetings return value to a customer.

6 Conclusion

This experiment indicates the potential of VR for allowing customers and bank employees to meet remotely. These meeting will have better engagement and focus if the results of this experiment can be replicated in the VR meeting software used. The hypothesis that participants would have more positive feelings to VR when compared to video communications (VC) meetings was only partially upheld and only in the case for female participants.

References

Campbell, A. G., Holz, T., Cosgrove, J., Harlick, M., & O'Sullivan, T. (2019). Uses of virtual reality for communication in financial services: A case study on comparing different telepresence interfaces: Virtual reality compared to video conferencing. In *Future of information and communication conference* (pp. 463–481). Springer.

Oprean, D., Simpson, M., & Klippel, A. (2018). Collaborating remotely: An evaluation of immersive capabilities on spatial experiences and team membership. *International Journal of Digital Earth, 11*(4), 420–436.

Schroeder, R., Heldal, I., & Tromp, J. (2006). The usability of collaborative virtual environments and methods for the analysis of interaction. *Presence: Teleoperators and Virtual Environments, 15*(6), 655–667.

How INVISTA's Transforming Manufacturing Training with Virtual Reality

Lou Pushelberg

Abstract

Can VR training improve manufacturing? It definitely can. In 2019, INVISTA came to Circuit Stream with a problem. Their current training model of slides and job shadowing was taking them months to train employees—six months actually.

What was more, once the employees were trained, they often still didn't have the hands-on experience needed for technical and skills-based roles. INVISTA was spending hours of manpower and losing valuable operational time while teaching new recruits. Plus, materials put toward training were thrown out; creating waste in the process.

This is where Circuit Stream came in. Circuit Stream applied its process for VR training to help INVISTA reduce costs while creating a faster, safer, and more engaging training experience. Read on to learn about the project's transformational results.

Keywords

VR · Virtual reality · Oculus quest · Enterprise · Training · Innovation · Manufacturing · Circuit stream · INVISTA

1 Company Description

Circuit Stream is a leading provider of augmented reality (AR) and virtual reality (VR) education, development and software services. Established in 2015, the company has built an ecosystem accelerating the adoption of AR/VR technology through

L. Pushelberg (✉)
Circuit Stream, Calgary, Canada
e-mail: lou@circuitstream.com

T. Jung, J. Dalton (eds.), *XR Case Studies*, Management for Professionals,
https://doi.org/10.1007/978-3-030-72781-9_16

three core pillars: (1) training courses, (2) software development services, and (3) an enterprise platform for scaling AR/VR applications.

The company serves the $6.3B AR/VR training market driven by large enterprises across multiple verticals. Through its educational and development services, Circuit Stream supports technical teams in developing the skills to build and use AR/VR technology. Its enterprise platform scales these solutions, solving common friction points and standardizing AR/VR learning protocols with analytic capabilities and demonstrable return on investment results. To date, the company has trained over 30,000 professionals across North America and has worked with some of the world's largest organizations including Koch Industries, VMWare, Global Foundries, Anglo American, Lockheed Martin and the US Navy helping them build internal skills and scale AR/VR adoption.

One of the company's Circuit Stream works closely with is INVISTA, one of the world's largest producers of chemical intermediates, polymers, and fibers. Headquartered in Wichita, Kansas, INVISTA has a presence in more than 20 countries. From the fibers in carpet to the plastic in automobiles, INVISTA's commitment to continuous improvement has led its employees to develop some of the most durable, versatile polymers and fibers in the world. A subsidiary of Koch Industries since 2004, INVISTA brings to market the proprietary ingredients for nylon 6,6 and recognized brands including STAINMASTER®, CORDURA® and ANTRON®. INVISTA also offers specialty chemical intermediates and process technologies.

2 Project Summary

Can virtual reality (VR) training improve manufacturing? It definitely can. In 2019, INVISTA came to Circuit Stream with a problem.

Their current model of traditional classroom training and job shadowing was taking them months to train groups of new employees, six months to be specific. Additionally, once the employees were trained, they still didn't have the hands-on experience needed for technical and skill-based roles. New groups of trainees, which were typically in the low double digits, were brought into the company multiple times throughout the year to conduct training for numerous operational roles, creating a significant opportunity for INVISTA to improve its technical training.

Using a traditional training model, INVISTA was spending months of manpower and losing valuable operational time, plus the materials used during training were often discarded, creating waste in the process. This is where Circuit Stream came in. Circuit Stream applied its process for VR training development to help INVISTA reduce costs while creating a faster, safer, and more engaging training experience. Read on to see the transformational results INVISTA and Circuit Stream experienced by implementing VR training in a manufacturing environment.

3 Project Details

INVISTA and Circuit Stream's work on virtual reality started when the site in Camden, South Carolina wanted to revamp its spinning technician training, a skilled role in which employees wind molten fiber around a complicated system of spools to stretch and strengthen it.

When performed correctly, the fiber is cooled through a combination of water and air and ultimately threaded onto several large spools. When the spools reach their final size and weight, they are removed from the machine and loaded onto a pallet where they will be shipped 'downstream' to another manufacturer. The spools of yarn will ultimately be used to make many everyday consumer products including things like carpet, car materials and even clothing.

But before the product gets into the hands of consumers, or even to the downstream manufacturer, INVISTA must ensure that their product is produced consistently, efficiently and most importantly safely, all results which stem from a highly effective training process.

String-up training itself can be difficult, time-consuming, costly and potentially dangerous for someone who is untrained. One of the biggest challenges in training new employees is that trainers had to wait for a filament to break before teaching a trainee how to restring it because there was no better way to practice than on the actual machine, and no training available that was similar to it. And in order to get the machine running again as soon as possible, training sessions lasted only 30 minutes. New employees needed six months on average to reach proficiency. "It really wasn't effective or efficient," one of the trainers, Robert Ray, said. "There was a lot of wasted time and frustration."

Dane Laughlin from INVISTA's innovation team sitting alongside some of the devices they are exploring and using. He shares more information about the VR training program in a video here: https://www.youtube.com/watch?v=vccyFPOOC4s

INVISTA's innovation team had been interested in virtual reality for a while, so when the Camden site asked if there were any ways to improve training, they jumped on the idea. A VR environment is more conducive to learning—instead of spending short bursts on a factory floor, trainees complete the simulation at a self-directed pace in a quiet room.

Repurposing one of the conference rooms, the Training team set up a permanent VR Training Center where new employees would engage with VR-based learning environments during their onboarding for many technical roles. With the goal of creating the highest fidelity experiences possible, the team installed several Valve Index VR headsets in the space, along with the corresponding hardware, and forecasted bringing dozens of new trainees through the center every year.

While the hardware was being installed, the next step was to design the experience. After extensive discussions with current employees, the team created a training simulation that incorporates visual cues as trainees "work" on the spinning machine. Check and information marks pop up to indicate what step comes next or warn that a task is incomplete. VR makes training less fraught because it eliminates the risks associated with untrained hands working on equipment using molten fiber and fast-moving components.

A user completes one of the steps in the VR simulation and is presented with the next task

It took about 90 days to develop the VR simulation, and there were some bumps along the way. At first, the team thought trainees should use an actual "string-up gun" while practicing with the VR headsets in the training room at the Camden facility. "But then we realized that four people swinging poles around in a facility training room while blindfolded wasn't a great idea," Dane Laughlin from the innovation team laughed (Koch Newsroom, 2020).

4 Feedback from End Users

The new system, which launched in November 2019, has the potential to cut training time in half. One huge benefit is that trainers can teach a whole group at once. It's also now possible to use one person's mistake as a teachable moment.

"It's so much more nuanced," Robert said. "You can show people that if they just tweak how they're moving their hand 10 degrees, it makes everything so much easier. That just wasn't possible when we were moving fast and yelling over the factory floor noise."

The training has had some unexpected benefits. The VR program turned training into a friendly competition, with employees wanting to be better and faster than their fellow trainees. That's in part because Circuit Stream's VR platform allows the company to capture detailed, real-time data on how well each trainee is mastering the work.

Data collected during training includes the completion rate by step, the average time per step and session, and the unique active users and sessions over time. Additionally, the Circuit Stream SDK (software development kit) and Platform collect and transmit more granular training data in the form of xAPI statements.

xAPI is a new e-learning specification that makes it possible to collect data across a range of technologies and experiences. A generalized example is that the xAPI system turns every learning behaviour into an actor (John), verb (turned), and object (the dial).

This new data format is great for interpreting simulations, such as AR/VR training, because it can measure specific behaviours and context. In the case of training for manufacturing equipment, this includes measuring a variety of metrics from within the simulation itself. Things like hazards (e.g. if an employee is too close to parts of the equipment that are extremely hot or spinning at a high velocity), what equipment the trainee is interacting with at each stage of the process, and even where the user's eyes are looking.

The outputs captured in the AR/VR application are expressed in xAPI formatted statements that are transmitted to a company's LRS (Learning Record Store), which is a specialized database conceived specifically to store and transmit xAPI data to other business systems, and ultimately consumed either in a LMS (Learning Management System) or other business intelligence tool.

This new way of capturing data enables vastly more insights for training managers and organizations to interpret how training is being performed and its effectiveness in more realistic and intuitive ways as compared to traditional methods.

"In the past, we had no objective way to measure a trainee's preparedness," said Adam Brooks, INVISTA IT business systems analyst. "Now, trainers can see where individual trainees are falling behind. They can aggregate data to determine which steps are more accident-prone, and take steps to mitigate the danger."

The VR training also has cut the time it takes for current employees to transition from one machine to another, so it's easier for them to switch jobs if they or the company need a change. Longtime employees have marveled at the VR training, with many joking new trainees "have it easy." Those trainees, meanwhile, have a hard time imagining learning any other way.

5 Future Outlook and Road Map

Next up for INVISTA's Innovation and Training team is an effort to scale VR training by putting it in the cloud, giving any site around the world access. Using a cloud-based approach coupled with a VR simulation is a prime example of INVISTA's goal to encourage employees to harness new technology to evolve their roles and processes for the better, said Steve Daley, president of Koch's Market-Based Management capability, which is responsible for cultivating an environment of continuous learning and transformation across Koch's worldwide facilities and nearly 130,000 employees. "Koch employees are using technology to facilitate learning and improved knowledge sharing to increase the rate at which we transform how work is done."

"This first VR program marks the beginning of a new chapter for INVISTA, now that the company can fully immerse a trainee," Brooks said. "We can run them through situations and scenarios, over and over again, without any risk and without any waste."

It's why the team now is developing more virtual reality training programs for other operational, health and safety tasks as well as various procedures throughout the company's many manufacturing facilities. INVISTA is also working on augmented reality for guided work instructions and remote assistance. In short, they want to turn as many costly and cumbersome processes into VR as they can to improve the employee experience. "We're looking at all sorts of other potential use cases," Dane said (Fig. 1).

6 Conclusion

In the end, the results from the VR training speak for themselves, employees are reaching proficiency up to 50% faster, total training time has been reduced by month.

Circuit Stream is developing more VR programs for INVISTA's manufacturing facilities and continues to focus on transforming the training process.

Meanwhile, Dane has been impressing spinning operators with his threading acumen. Recently, while visiting a site, he was challenged to thread a live machine, something he'd never attempted. "It was kind of funny because the whole way, the

Fig. 1 With its first virtual reality training program successfully launched, INVISTA's Innovation team is seeking to scale VR-based training to sites globally. The team is also exploring the use of augmented reality (AR) for providing work instructions and remote assistance. Dane Laughlin is seen above wearing a Microsoft Hololens augmented reality headset

operator was telling me how hard it is to catch the thread to get it started," Dane said. "I was kind of worried. But I had a very strange sense of déjà vu, because I'd never actually done it outside of VR, but I felt like I had. And then it was snip, snip, done! His jaw just dropped. That was pretty cool."

Reference

Koch Newsroom. (2020). *Too cool for spool: New virtual reality system transforms INVISTA training regimen*. Wichita.

Use of Virtual Reality to Support Rapid Upskilling of Healthcare Professionals during COVID-19 Pandemic

Alex Young and Alex Aquilina

Abstract

During the first six months of the COVID-19 crisis Virti, a UK based training and development company, created, distributed, and analysed training and development support to NHS organisations across the United Kingdom and United States. This innovative and rapid response to the emerging needs of the healthcare sector has demonstrated how VR can be used to provide adaptive training at scale and pace.

1 Company Description

Virti is a digital training company pioneering the use of XR, artificial intelligence (AI) and gamification to help employees learn faster and retain knowledge for longer.

Founded in 2018 by NHS Trauma and Orthopedic surgeon Dr. Alexander Young, Virti makes training scalable by allowing learners to access immersive and interactive training on any headset or smartphone, and allows organisations to create their own educational content.

A. Young (✉) · A. Aquilina
Virti Inc, San Mateo, USA
e-mail: alex@virti.com; alex.aquilina@nhs.net

T. Jung, J. Dalton (eds.), *XR Case Studies*, Management for Professionals,
https://doi.org/10.1007/978-3-030-72781-9_17

A user runs through Virti training on their smartphone

Virti believes that on-demand, just-in-time training enabled by XR technology can generate superior educational outcomes in a zero-risk environment. What's more, the unique objective data captured via AI can provide unparalleled pedagogical insights for learners and trainers.

Virti is the first and only evidence-based training company to be selected onto the prestigious NHS Innovation Accelerator programme and deployed in the UK's NHS.

The company ranked highest in the 2017/18 Digital Health. London accelerator cohort, is part of the Texas Medical Center's innovation programme TMCx in Houston, received the Royal College of Surgeons of Edinburgh Triennial Innovation Prize and topped the VR Healthcare Category at the VR Awards 2018.

2 Project Summary

During the COVID-19 pandemic health providers struggled on a global scale to meet the demands of increased patient volumes. In order to meet demand, providers turned to volunteers and retired health professionals in addition to reassigning existing clinical staff and medical students to specialties and areas of practise with which they were not immediately familiar such as intensive care.

In the community carers and volunteers are at an equally increased risk, albeit with less access to training, on how to use personal protective equipment correctly. The problem providers face is how to quickly train staff at scale so that they are safe and feel prepared to meet the stresses imposed on the system by COVID-19.

During the COVID-19 crisis, Virti worked with the UK's NHS, one of the largest health providers globally employing more than 1.5 million people, together with health providers in the US and internationally.

To give an idea of the extent of the training required during COVID-19, in the state of New York alone more than 40,000 healthcare workers, including retirees and students, volunteered to be part of New York's surge healthcare force. In the UK over half a million volunteers signed up to support clinical staff with Health Secretary Matt Hancock announcing that 11,788 retired medical staffers would return to the National Health Service, including 2660 doctors and 6147 nurses. In addition, 5500 final year medical students and 18,000 nursing students were moved onto the front line with many graduating early to help out.

With access to face-to-face training drastically limited—and the need to equip frontline healthcare workers with COVID-specific skills critical, Virti stepped in to deliver remote educational programmes to NHS employees at scale.

Virti placed healthcare professionals including doctors, nurses and carers into on-demand, experiential training environments from the safety of their mobile device or immersive headsets. Scalable, repetitive practice combined with an increased sense of presence maximises learning effectiveness in otherwise serendipitous or hard to access face-to-face training. Unique data-driven insights then provide objective data on employee performance which would usually be subjectively assessed or missed during real world training opportunities.

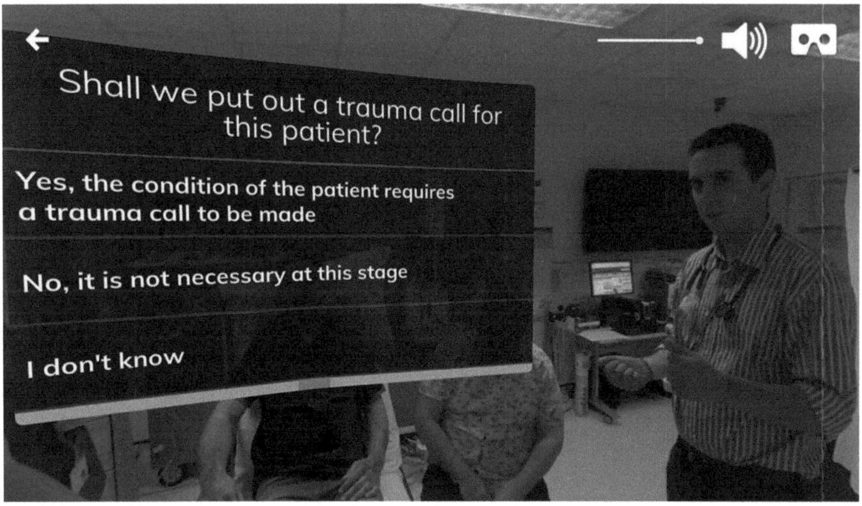

A screenshot during a segment of the training where the user has to make a decision

In response to the COVID-19 pandemic, Virti launched Virti-COVID to help frontline healthcare staff and carers quickly train across a wide range of areas from how to safely put on and take off protective equipment to using ventilators to having difficult end-of-life conversations with patients and families.

Doctors and nurses found it hard to communicate with patients while wearing protective gear, which could unintentionally alarm or intimidate them. One of Virti's simulations puts trainees in the position of a coronavirus patient, surrounded by medical staff in masks and gowns to help develop empathy and understanding by embodying the learner in the role of the patient.

Following approval by Health Education England, Virti was able to roll out specially designed COVID-19 modules for use on their immersive training platform—accessible to NHS staff via a virtual reality headset, desktop or mobile device.

Virti rolled out COVID-19 training for free to the NHS and also in the United States working with the Nightingale Field Hospitals and health providers. Virti also made the platform and training modules available to carers and volunteers. For this Virti won the NHS COVID-19 Challenge and were featured on the Nasdaq Tower in Times Square and in national press in the UK and US. Virti then received funding from NHSx to conduct a randomised control trial to analyse the impact of the COVID-19 training that showed the system could improve understanding of infection control measures amongst carers by 76% and improve their knowledge retention of crucial health and safety guidelines by 230%.

Internationally, Virti deployed COVID-19 training across large healthcare providers in the United States including Cedars-Sinai Medical Center and in resource limited environments including Ethiopia where access to medical training was limited.

Virti was featured as part of the Department for International Trade's GREAT Britain Campaign highlighting inspirational businesses who provided help and support during the COVID-19 pandemic.

3 Project Details

As part of the roll-out Virti was awarded funding by the NHS to conduct a feasibility trial to ascertain whether immersive training improved learning outcomes and performance for a cohort of healthcare professionals, volunteers and carers compared with traditional methods of teaching and training delivered to this group. Secondary outcomes measured include improved stress and anxiety and Net Promoter Scores (NPS) as well as platform usage and qualitative feedback.

The non-blind, randomised control interventional study recruited 50 healthcare professionals, volunteers and carers who had not previously received COVID-19 surge response training. The pre and post-training test focused on knowledge retention of the key steps in each scenario such as donning/doffing Personal Protective Equipment (PPE), hand washing and CPR, mapped to the steps set out in national guidelines.

A statistically significant reduction ($p < 0.05$) was seen in pre and post-training anxiety scores when comparing the intervention and control groups both overall and by sex. The intervention group showed a reduction in COVID-related anxiety levels and stress.

On average the intervention group saw a percentage increase in performance from pre to post-training knowledge test of 230.1% compared to 16.75% for the control group indicating significantly improved retention and application of training knowledge. Based on test scores, 42% of the control group were deemed to have adequate understanding of infection control measures compared to 92% of the intervention group.

Number needed to treat (NNT) was calculated as 1.32 demonstrating strong effectiveness of the intervention over traditional training methods.

A health economics study revealed estimated savings of £1898.6 per employee per year on the cost of training and savings of £640.84 per employee on loss of productivity and absenteeism due to COVID-infection and stress.

Further cost impacts include reduction in carer/HCP family infection and hospitalisation—the costs associated with related patient infection and deaths are more challenging to estimate given the time frame of the study and the potential confounding factors such as availability of PPE and underlying health issues of the HCPs, carers and patients.

The study met its aim in demonstrating that immersive training can rapidly upskill the health and social care workforce quickly and effectively and can also reduce anxiety in this group of employees.

4 Feedback from End Users

Tom Woollard, West Suffolk Hospital Clinical Skills and Simulation Tutor, commented, "We've been using Virti's technology in our intensive care unit to help train staff who have been drafted in to deal with COVID-19 demand. Training is being accessed on the Virti platform by nursing staff, physiotherapists and Operational Department Practitioners (ODPs) to orient them in the new environment and reduce their anxiety. The tech has helped us to reach a large audience and deliver formerly labour-intensive training and teaching which is now impossible with social distancing."

Russell Metcalfe-Smith, Director of the Women's Guild Simulation Center for Advanced Clinical Skills at Cedars-Sinai Medical Center has been using Virti across a range of a wide range of workforce problems: orienting employees to a new department, surgical training, communication skills, bullying and harassment compliance training, and even mindfulness and resilience training for staff. Most recently he has been using it for COVID-19 surge response training.

"We've found it very valuable to observe a doctor's thinking process and it's all been done with social distancing. Because we can't get large groups together right now, we've had to rely on technology to give the same experience. The simulations provide a feeling of presence and it feels like you are in a room with a patient. Based on the decisions you make one direction will lead to another. We have doctors jump quickly into a virtual environment like this to get them to where they need to be."

Dr. Alec Snow, an intensive care and anaesthetic registrar at the Royal United Hospitals trust in Bath, used Virti COVID training during the pandemic. He said that his hospital had been conducting face-to-face training "non-stop" for the first few months of COVID-19.

"A lot of us cancelled annual leave, and I worked through my parental leave. There was a huge appetite and willingness from other staff groups to help, but the skills between specialisms are vastly different, and the skills that were needed for managing COVID-positive patients are quite specialised in themselves, so fast and impactful training is needed. Prior to using Virti's immersive training, we were resorting to recording simulation training videos and sharing via Whatsapp which was not very engaging nor very helpful, so the Virti training has been incredibly useful at keeping us updated and alleviating any anxiety. The real challenge with the pandemic was maintaining safety for everyone and not cross-infecting staff. If we had lost our staff then our ability to manage the outbreak would have been extremely hampered—this is where interactive immersive technology can have an exponential impact on not just staff but the whole NHS."

5 Future Outlook/Roadmap

The use of the Virti platform as a training tool to train healthcare professionals and reduce anxiety is one of the many uses of immersive technology in healthcare education. As this study suggests, immersive technology can potentially be used to improve non-clinically trained staff in a short space of time. However, more research is required. The use of Virti as a teaching tool is not just limited to teaching COVID-19-related skills. There are many potential use cases in health and social care beyond these scenarios including accelerating care certification, refresher and re-certification training, and more soft skill based modalities including but not limited to: roles and responsibilities; whistleblowing; equality, diversity and person centred values; telephone support and communication; pain and discomfort; infection prevention and control; food safety; fluids and hydration; health and safety and stress management.

The exponential use of immersive technology in simulation has significantly advanced provider education. Simulation has been recognized for its approach in "...[improving] the acquisition of knowledge and skills and [strengthening] quality and safety in clinical practice." [(30)] From software applications to virtual avatars, evidence-based research has shown immersive technology can strengthen physician and nursing training.

Immersive technology can be integrated into traditional simulation pre-sim, intra-sim and post-sim. In the pre-sim setting immersive tech can help to scale simulation training such that learners are better prepared for the time they spend in high-fidelity in-person simulation and use this time to hone skills rather than to learn them for the first time. Intra-simulation immersive technology can help to improve through-flow when learner numbers are high by using headsets outside of designated sim time. Finally, skill-fade can be reduced post-sim by allowing learners convenience of access to repeat the simulations in VR either in headsets or via mobile.

Effective educational practice will help healthcare professionals to understand the relationship between theory and clinical practice, which will enable them to exercise better judgement in complex situations. They will also be encouraged to understand other roles within the team and show how they can adapt and collaborate in emergency situations. HCPs will need to become aware of the different perspectives and expertise that can improve problem solving, clinical reasoning, patient management and decision-making. This depth of knowledge, applied skills, practical skills, soft skills and judgement is best represented by the metaphor of the iceberg.

Immersive technology compliments this learner growth by providing convenient, on-demand access to a wide variety of learning opportunities from which unique data insights can be derived to complement reflective feedback and maximise performance. Immersive offers a wide variety of learning opportunities beyond simply scaling traditional simulation training.

6 Conclusion

The project has demonstrated that immersive training can rapidly upskill the healthcare workforce quickly and effectively and can also reduce anxiety in this group of employees through improved training.

The project considered the efficacy of using a remote, immersive training platform to reduce anxiety and improve clinical decision making and knowledge retention for health professionals who needed to be upskilled quickly in response to the COVID-19 pandemic.

Participants who had access to the intervention performed better in their test with a higher mean score in the post test and a higher numbers of healthcare professionals deemed to have adequate knowledge of COVID-19 infection control methods. Intervention participants also reported more confidence in their knowledge and application of knowledge with a reduction in COVID-related stress and anxiety.

Knowledge retention was significantly higher in the intervention group suggesting that important clinical information can be deployed and retained by users in a short space of time using the intervention method. Improved training corresponded with reduced anxiety scores indicating a potential benefit for employee stress and well-being. The questionnaire results demonstrate that participants in the intervention group had greater confidence when asked if they knew the steps of tasks as well as confidence to perform tasks in real life. Participants also enjoyed the immersive teaching course compared to traditional methods of learning and this was supported through the platform usage data.

The health economics data revealed significant savings both to the direct cost of training and to improved return on investment.

References

Berger, M. L., Murray, J. F., Xu, J., & Pauly, M. (2001). Alternative valuations of work loss and productivity. *Journal of Occupational and Environmental Medicine, 43*(1), 18–24.

Bowman, D. A. et al. (2009). Higher levels of immersion improve procedure memorization performance.

Deloitte. (2014). *Mental health and employers: The case for investment*. https://www2.deloitte.com/uk/en/pages/publicsector/articles/mental-health-employersreview.html

Gutiérrez, F., Pierce, J., Vergara, V. M., et al. (2007). The effect of degree of immersion upon learning performance in virtual reality simulations for medical education. *Studies in Health Technology and Informatics, 125*, 155–160.

http://usatoday30.usatoday.com/news/nation/2004-03-01-crash_x.htm

https://digital.nhs.uk/data-and-information/publications/statistical/nhs-workforce-statistics/february-2020

https://hbr.org/2016/10/why-leadership-training-fails-and-what-to-do-about-it

https://trainingmag.com/trgmag-article/2019-training-industry-report/

https://www.ft.com/content/46baab28-6def-11ea-89df-41bea055720b

https://www.gartner.com/en/human-resources/role/learning-development

https://www.gov.uk/government/publications/covid-19-guidance-on-social-distancing-and-for-vulnerable-people/guidance-on-social-distancing-for-everyone-in-the-uk-and-protecting-older-people-and-vulnerable-adults#how-do-you-look-after-your-mental-wellbeing

https://www.hee.nhs.uk/sites/default/files/documents/Workforce%20Stress%20and%20the%20Supportive%20Organisation_0.pdf

https://www.mckinsey.com/business-functions/organization/our-insights/getting-more-from-your-training-programs

https://www.nuffieldtrust.org.uk/resource/the-nhs-workforce-in-numbers

https://www.nytimes.com/2020/03/31/climate/coronavirus-doctors-retire.html

https://www.resus.org.uk/media/statements/resuscitation-council-uk-statements-on-covid-19-coronavirus-cpr-and-resuscitation/

https://www.shiftelearning.com/blog/statistics-on-corporate-training-and-what-they-mean-for-your-companys-future

https://www.telegraph.co.uk/global-health/science-and-disease/coronavirus-news-uk-schools-cases-deaths-london-lockdown/

https://www.theguardian.com/world/2020/mar/25/astonishing-170000-people-sign-up-to-be-nhs-volunteers-in-15-hours-coronavirus

Iglesias-Vázquez, J. (2007). Cost-efficiency assessment of advanced life support (ALS) courses based on the comparison of advanced simulators with conventional manikins. *BMC Emergency Medicine, 7*, 18.

Issenberg, S. B., McGaghie, W. C., Petrusa, E. R., Lee Gordon, D., & Scalese, R. J. (2005). Features and uses of high-fidelity medical simulations that lead to effective learning: A BEME systematic review. *Medical Teacher, 27*, 10–28.

Mattke, S., Balakrishnan, A., Bergamo, G., & Newberry, S. J. (2007). A review of methods to measure health-related productivity loss. *Am J Managed Care., 13*, 211–217.

McIntish, C. (2006). Simulation: What does it really cost? *Simulation in Healthcare, 1*, 109.

Memory: *A contribution to experimental psychology—Ebbinghaus* (1885/1913)".

Nicholson, S., Pauly, M. V., Polsky, D., Sharda, C., Szrek, H., & Berger, M. L. (2006). Measuring the effects of work loss on productivity with team production. *Health Economics, 15*, 111–123.

Padilha, J. M., Machado, P. P., Ribeiro, A., Ramos, J., & Costa, P. (2018). Clinical virtual simulation in nursing education: Randomized controlled trial (Preprint). doi:https://doi.org/10.2196/preprints.11529.

Repetto, C., Serino, S., Macedonia, M., & Riva, G. (2016). Virtual reality as an embodied tool to enhance episodic memory in elderly. *Frontiers in Psychology, 7*, 1839. https://doi.org/10.3389/fpsyg.2016.01839

United States Department of Labor. (2008). *Employer costs for employee compensation: December 2008*. U.S. Bureau of Labor Statistics.

Vozenilek, J., Huff, J. S., Reznek, M., & James, A. (2004). See one, do one, teach one: Advanced technology in medical education. *Gordon Acad Emerg Med, 11*(11), 1149–1154. https://doi.org/10.1197/j.aem.2004.08.003

Walsh, K., & Jaye, P. (2013). Simulation-based medical education: Cost measurement must be comprehensive. *Surgery, 153*, 302.

Zendejas, B. (2013). Cost: The missing outcome in simulation-based medical education research: A systematic review. *Surgery, 153*, 160–176.

Cost-Effective and Eco-Friendly Fire Investigation Training Using Photorealistic Interactive Room Scale Virtual Reality

Bradley Friend Woodward

Abstract

Reality in Virtual Reality Limited (RiVR) conducted research with various fire services in UK, USA, and Europe to understand how fire departments train personnel in the discipline of fire investigation. The aim was to identify if this task could be replicated in VR using an immersive VR application to improve the health and safety of practitioners, reduce the ecological impact, save costs and improve accessibility. Although the product has only been released in 2020, the feedback and adoption provides a good example of the efficacy of immersive content for training and development in this particular field and has also helped to create a roadmap for future expansion across the blue light service community.

Keywords

Photo realistic · Interactive · Replicable · Virtual reality; VRM · Photogrammetry

1 Company Overview

RiVR is a UK company created in 2017. It is a specialist team of video game developers, software engineers, digital artists, drone pilots, ex-service personnel, cinematographers, and trainers. The organisation is a developer of virtual reality assets in both 360-degree video, stereoscopic video (3D) and photorealistic virtual reality experiences. It offers immersive training & experiences for all industries.

One area that RiVR has become synonymous with is the scanning of any real-world environment and the use of photogrammetry to create a virtual twin of any real-world space in a photorealistic environment. These environments allow the use

B. F. Woodward (✉)
Reality in Virtual Reality Limited, Southam, UK
e-mail: Bradley.woodward@rivr.uk

T. Jung, J. Dalton (eds.), *XR Case Studies*, Management for Professionals,
https://doi.org/10.1007/978-3-030-72781-9_18

of VR simulation to activate special memory around a situation which stimulates muscle memory and skills often taught in hazardous situations. RiVR allows users to interact with and experience these worlds; thus, enhancing the way humans learn.

One of its most prestigious contracts is with Transport for London (TfL) supporting the training of 28,500 London bus drivers through the ten bus operator companies using bespoke RiVR Link software (a synchronised instructor platform for group training).

2 Project Summary

RiVR in January 2020 launched 'RiVR Investigate'. This is a virtual reality photorealistic room scale interactive training package specifically for the training of fire investigation. It comprises a library of post fire scenes and is sold under an annual license fee. Overall, the aim is to create a programme that could not only replicate the way in which training is currently being conducted globally, but also to add additional features, provide consistency to experiences and massively reduce the costs that fire departments are having to withstand.

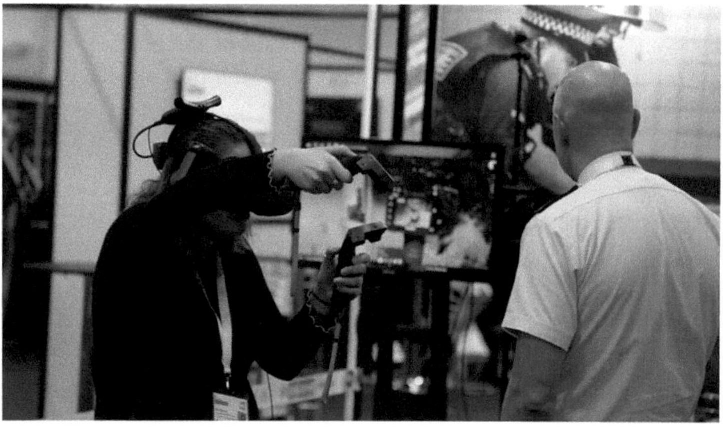

RiVR Investigate training in action

In addition, RiVR requested The Life & Sciences department of Coventry University to conduct independent research on the efficacy of the application, using Leicestershire Fire and Rescue Service as the test organisation. Due to COVID-19. quantitative data is limited as training was halted in most institutions. There is no doubt that there is an appetite for the product with adoption in Finland, Far East, Australasia, USA, and Europe.

3 Project Details

3.1 The Situation

The purpose of fire investigation training is for fire fighters to establish the origin of the fire, determine the likely cause and to arrive at a conclusion as to whether the event was accidental, natural, or deliberate. The establishment of causation is to reduce similar events occurring in the future (if the case is natural or accidental) or to allow a legal investigation to be conducted (in the case of a deliberate fire).

There is a global method of training fire investigators. This consists of classroom studies followed by a 'controlled burn'. Most fire departments use ISO (shipping) containers to create this experience. They 'dress' the container using plasterboard/plywood; adding objects to make a realistic room. They then use a source of ignition to create the fire. The fire is allowed to burn to a certain level and is the extinguished. After a period, allowing the ash to settle, students are then invited to study the scene, using the classroom and previously obtained knowledge. They make contemporaneous notes, use evidence markers, and take photographs to identify the source of ignition. Exploration is limited as the scene must be preserved for subsequent students to go through.

This real-world simulation is to ensure that training is safely conducted.

Until now, this has been the most cost-effective way of training but does have limitations.

- Time consuming to create these experiences (obtaining 'props', building the scene, etc)
- Not eco-friendly (burning items and then the need to dispose of remains)
- Expensive (cost of props, time required to create, student hours)
- Heath risk to fire fighters (any fire, even in controlled environment carries risk)
- Health risk to instructors & students (carcinogens being absorbed through the skin enhancing risk of cancer)
- Limited exposure for each student (unable to truly examine as not allowed to disturb the scene as the next student would not get the same experience)
- Time bound (little flexibility in terms of when the training can be held)
- Students unable to see, in retrospect, what caused the fire.
- Students unable to utilise the services of a trained sniffer dog when they suspect that an accelerant has been used (A limited resource that is not generally available in training establishments.)
- Many departments do not have access to create their own 'burns' due to location and cost (they send their staff to other services/organisations that run such sessions at considerable cost)

Traditional method of dressing a burn scene

Due to the limitations of the current training, fire departments are keen to look at the art of the possible in terms of creating a more mobile and agile system of training. This is particularly driven in the UK where a change in national operational guidelines indicated that a significantly larger population of fire fighters would have to undergo level 1 fire investigation training. RiVR was supported in this process by Leicestershire Fire and Rescue Service who initially identified this method of training as an opportunity that might be appropriate for a virtual reality application.

The Solution

During 2018/2019, the company embarked on a full programme of research to understand current practices in the fire service industry and to design a repeatable training programme that can accredit an individual's performance to the same standard.

- **Review of UK National Operational Guidance**—Understanding the impact in the UK of planned changes in fire service guidelines. RiVR aimed to create a training package that not only embraces these guidelines but, using new technology, also improves learning outcomes by recording all processes that can be replayed at any point in the future.
- **Review of overseas current practices**—Visits to FLETC (Federal Law Enforcement Center) in USA and Civil Defence Academy Singapore to identify any significantly different methodology.
- **Proof of concept scene**—Exhibited at exhibitions and expos in USA, France, Sweden, Finland, Holland, UAE to gauge market appetite and potential efficacy.

Following this undertaking, RiVR interviewed a number of UK fire investigation experts to identify a person or organisation that could assist in the design of the learning outcomes. After a lengthy process, Jason Dean MSc, MIFireE, FFireInv, MCSFS, was appointed consultant to design the library scenes and to ensure that the realistic scenes being created by RiVR were of training value and replicated real historic events.

RiVR purchased two ISO containers and established a safe environment on a farm to carry out 6 controlled burns over a 10-month period. The scenes were designed and created to a detailed specification.

Container scene post burn

Prior to any event taking place the process of photogrammetry was used to 'capture' the scene. A source of ignition was used in line with the scene 'backstory'. The burn was then captured using a 360-video bespoke camera rig (designed inhouse). At an appropriate time during the burns, the fire was extinguished using professional firefighters. At this point in the process a fire service sniffer dog was introduced into the container and a 360-video captured to show whether the dog indicated that an accelerant had been used. The scenes were then captured again using laser scanning data and thousands of photographs. The final part of asset capture entailed each single burnt object within the scene being meticulously removed and scanned using a specially designed RiVR camera rig.

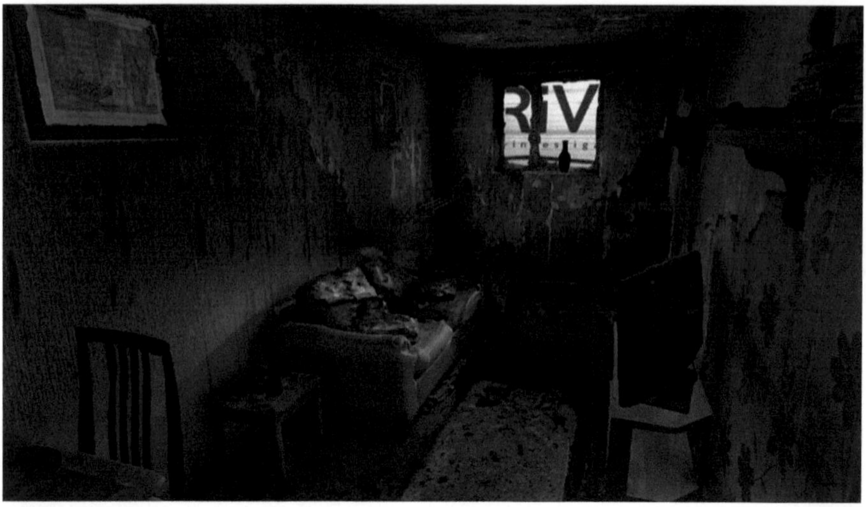

Digital reconstruction of burn scene

Each scene was then recreated using various software applications to create a digital twin of the original environments. All the digital objects were then 'placed' in a photorealistic forensic laboratory allowing for examination with several user tools.

Details of RiVR Investigate System

RiVR software application was built for the HTC Vive VR headset as an all-in-one immersive technology solution. The bespoke software allows user mobility around the scene and total object interaction along with the following features.

- Instructor interface (Virtual Reality Monitor)
- Evidence single objects/ Evidence multiple objects
- Detailed Examination
- Take photographs (with the ability to print them in the real world)
- Torchlight (for close up analysis, as in the real world)
- Mobile beam lights (Suspended lights for clear viewing)
- Dictaphone (recording of contemporaneous notes)
- Instructor viewing (first person/ third person/bird's eye view)
- Multi-user functionality
- Communications via headset
- Cloud based recording
- Ability to replay all activity of the student in the scene from any view
- Student interactions (students can request a sniffer dog and, at the end of the training, view how the fire started)

3.2 Multi-User Functionality

Multi-user functionality was always part of the original specification for RiVR Investigate but the need for this was accelerated by the COVID-19 pandemic due to social distancing and has now become an integral feature of the system.

3.3 Evaluating the Impact

The product has been purchased by fire departments in USA, Australia, Singapore, and Europe. Each department has managed to quantify cost savings but due to the global nature and variances in working practices these savings are inconsistent in monetary terms alone. However, the cost of a Fire Service creating six real world burns is in excess of £36,000. The RiVR library of six scenes is purchased under licence at a cost of £15,000 with additional scenes being added annually at no additional cost. Other than financial benefits, additional benefits are being recorded in terms of training being more ecologically friendly, replicable, consistent, mobile, and recordable.

Fire Investigator examining a RiVR scene

The outcome from the research being conducted by Coventry University will provide additional data on the efficacy of the solution. This report has been delayed due to COVID-19 and is anticipated to be published first quarter of 2022.

4 Feedback from End Users

The launch of the product has led to extremely positive feedback. Below are recorded comments from Fire Investigator trainers from WS Darley USA, Leicestershire Fire and Rescue Service UK, West Midlands Fire and Rescue Service UK, Research and Development Office, Royal Netherlands Army, Fire Safety and Investigation Branch, Civil Defence Academy, Singapore.

Question areas	Indicative responses
What did you think of RiVR investigate?	*"I can't believe how real it is. I actually thought I was there. It could save us a fortune in travel costs alone."* *"I must admit that I was sceptical, but it really is amazing. I never thought that we would be able to train fire investigation in virtual reality. It really is amazing and the detail is extraordinary."* *"Almost immediately I found myself going through the routine that I would in the real world. To see how the fire actually started was a game changer for me. Awesome!"* *"Simply outstanding. It is so real that I brushed my trouser after kneeling down to get rid of the ash!"*
How has RiVR investigate helped training in your department?	*"It gives our students the chance to operate in multiple scenes without the hassle of having to keep setting up and burning stuff!"* *"At the press of a button I can go again. It does everything that we need it to do—And more."* *"It's amazing to be able to debrief [with the students] and be back in the scene with them again. The tool kit is pretty mind blowing and it's so immersive. Students immediately embrace the technology."* *"It's safe, mobile and we can tick every learning outcome box."* *"Unbelievably immersive and we can create case studies better than we can in the real world. VR dog—Amazing!"*
Comparison with traditional real-world training methods	*"It's very engaging and creates shared experiences when we cast to a big screen, so everyone benefits."* *"It means that every student gets to truly examine the scene without risk of disturbing it for the next one."* *"Some of our younger fire fighters have been brought up using virtual reality and no period of adjustment is required"* *"It's cleaner, safer and quicker with no negative impact on the learning outcomes."* *"I think it will completely change the way in which we do a lot of our training in the future. I'm one hundred percent converted."* *"For the first time we have a solution that allows you to do more in the virtual word than we can teach in the real one. And it saves money."*

5 Outlook/Roadmap

RiVR will add two additional fire scenarios to the library each year as a minimum to ensure longevity of the product. There is the option for individual fire departments to create their own burn scenes, where RiVR will attend their location and use photogrammetry to create user bespoke content. This is a chargeable event where a financial discount is offered if the content can be added to the library.

A RiVR Investigate user group will be created in 2022 which will allow instructors to meet in a virtual environment to discuss future developments/ improvements and training use cases. This user community programme will be supported by a quarterly newsletter.

There is also a research programme now in place to allow the assets to be used in standalone headsets to allow further mobility without the need to be tethered to a PC.

Following the success of RiVR Investigate for fire investigation training, a proof of concept was developed for crime scene investigation. This is a sandbox solution to aid instructors in creating their own photorealistic crime scenes for forensic investigation from a large library of assets (including locations, objects, and bodies). This will be set to market in the first quarter of 2022.

6 Conclusion

Fire investigation is expensive for organisations to manage for trainees and is not without safety risk, not eco-friendly, and has limitations in respect of recreating a realistic environment for each individual trainee. RiVR Investigate forms part of a blended approach to learning outcomes in this discipline that is effective.

There are five key factors that contribute to the success of RiVR's VR fire investigation training solution.

Success Factor #1: A Solution to an Existing Problem

Replicable training is now provided where the environment is consistent for all trainees who also benefit from being able to fully investigate the scenes and catalogue their findings as they would in the real world. Because the scenes are photo-realistic, retention of information is accelerated and reinforced by instructors being able to offer tutorial guidance in the scene using the multi-user option. RiVR Investigate was designed to be a mobile solution where the trainee and the instructor do not have to be in the same location but can both be in VR at the same time anywhere in the world. This flexibility has the potential to significantly reduce costs.

Success Factor #2: Learning Outcomes

Each scene is recorded from one of three viewpoints and the trainee can, at the end of their analysis, discover if their findings are correct in line with how the original fire started. The recordability of the actions of the attendee allows for a personal debriefing process where they can analyse their performance to understand learning outcomes in a safe replicable way, which can also be reviewed at a later date to monitor progress.

The student is also able to present a case study for accreditation should the relevant fire service decide to use the system as an accreditation tool.

RiVR eagerly awaits the independent research paper being completed by Coventry University comparing learning outcomes from VR versus the traditional way in which fire investigation is currently being conducted.

Success Factor #3: Community Sharing Platform
The fire investigation community is, in relative terms, a small one globally. RiVR Investigate's product allows these individual specialists to come together in an immersive way to share ideas, improve core skills and become instrumental in the way in which the product continues to be developed.

Success Factor #4: Integration into Wider Applications
Successful delivery of this initial product has created an appetite in the forensic community for a deliverable CSI library where initially there is a 'cross over' into fire scenes. Development of the software engine that processes the assets created means that RiVR can create RiVR CSI in a relatively short time scale.

Success Factor #5: A Long-Term Strategy
RiVR has a commitment to enable post event analysis in virtual reality for the blue light services and other applicable industries including traditional seats of learning. It not only uses photogrammetry to create realism but has also embraced additional technology solutions such as 360 video content, LiDAR scanning, point cloud data capture, motion capture and analytical data capture to continually review how the products being created can be improved.

Being at the forefront of the art of the possible in immersive technology is at the core of the ethos of RiVR whilst always being mindful that successful outcome of the training process is the key component to adoption.

The company has an exciting roadmap to engage with end users in a multitude of industries and is committed to enhancing the process that humans undertake to develop new knowledge and skills in a photorealistic replication of the real world.

Reuniting Hospitalised and Isolated People with their Beloved Ones outside the Hospital with the Help of VR Livestreaming

VRiend a Stand-Alone, Low Latency VR Streaming Platform

Bas Beukers

Abstract

Horus VR developed a low latency VR livestreaming platform called VRiend. VRiend enables patients and isolated people to reconnect and communicate with their beloved ones outside the hospital. Patients could go back to school and follow lessons again, attend business meeting at work to stay up to speed with their colleagues or just have a chat with their grandchildren back home.

Keywords

Virtual reality · Live streaming · Do-it-yourself · Low latency · Construction · Connected, innovation

1 Company Description

Horus VR has used its extensive experience in Virtual Reality (VR) design and production to develop a "do-it-yourself" VR platform and tool-set which allows corporate customers to create, enrich and distribute VR content and interactive e-commerce via low-latency streaming or on-demand.

Their focus is to supply their clients with proven, user friendly and affordable VR ecosystems to create impactful VR experiences themselves. Horus VR as part of the Horus Holding employs 30 people and is based in The Netherlands, headquartered in Groningen.

B. Beukers (✉)
Horus VR, Groningen, The Netherlands
e-mail: b.beukers@horus-vr.com

T. Jung, J. Dalton (eds.), *XR Case Studies*, Management for Professionals,
https://doi.org/10.1007/978-3-030-72781-9_19

1.1 Markets & Solutions

Horus VR has targeted two market sectors: Healthcare and Entertainment. The approach has been to establish paying proof-of-concept customers in each sector, aiming to establish the de-facto standard prior to rolling out sector solutions more widely in conjunction with resellers and distribution partners.

The Company has succeeded thus far in attracting several anchor customers across each of its two sectors healthcare and entertainment. For now the main focus and core product is VRiend in the healthcare industry.

VR-Macy

Besides our VRiend we built the VR-Macy application. VR-Macy is our VR video library containing a huge amount of 360 video and tailored with the help and knowledge of professionals in the Healthcare industry into VR Health experiences to relax, distract, train and entertain patients. Our VR library includes also an ergometer training program during long lasting medical treatments like hemodialysis (Fig. 1).

2 Project Summary

Horus VR developed a low latency VR livestreaming platform enabling patients to reconnect and communicate with their beloved ones outside the hospital. Patients could go back to school and follow lessons again, attend business meeting and stay

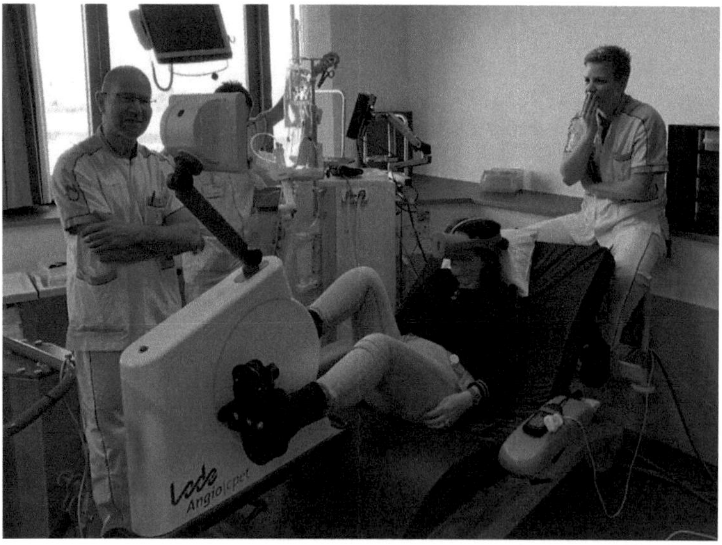

Fig. 1 VR Ergometer experience Deventer Hospital

up to speed with their colleagues *or just have a chat with their grandchildren back home.*

3 Project Details

3.1 Ambition

Horus VR started with the VRiend adventure end of 2015. Horus VR's ambition was to reunite hospitalised and isolated people with their families during Christmas. Our first test case was a huge success and proved this set up good work to reunite people.

3.2 Challenges

We started the VRiend in 2015 with high end expensive camera equipment operated by a professional crew at both sides. This to set up, control and to configurate the livestream on both ends. The experience was great but not scalable, a technical challenge and 30 seconds of latency preventing you to communicate in real time.

End of 2016 we finally had our first MVP. With a custom-built camera, an Android play back player on a Samsung GearVR and a P2P streaming platform in the middle we were able to test the basics of VR low latency VR streaming.

In 2017 we built the playback app for the new standalone VR Head Mounted Displays [HMD] and we designed a new housing of the camera. This was the first version allowing you to set up a VR livestream yourself.

In 2018 we did a major redesign of the VRiend housing and we completely rebuilt the *printed circuit board [*PCB] and the inside of VRiend. This to decrease latency, a new power supply to increase period of use, get more feedback of the system, make it more stable and user friendly. VRiend 1.0 was officially launched in 2018 Q2 and we started shipping the first batch of 20 VRiends.

4 VRiend Framework

Horus VR developed a custom VR camera device, a bespoke low-latency streaming platform including a play-back app. Enables us to get full control, a secure connection and low latency feedback instead of using common commercial platforms like Youtube and Facebook.

4.1 VRiend Camera

The VRiend streaming camera has a dome shaped lens on top his head. This wide-angle lens covers 280 degrees of field of view. Inside VRiend you will find a 4G dongle and a Wi-Fi router to set up your connection. The 2 batteries power supply

Fig. 2 VRiend camera system including the Oculus GO HMD

will last for 2 *hours of livestreaming. Before you need to charge VRiend. The custom designed PCB is the core of VRiend by running the OS, being the streaming server and give you instant feedback. The sound is managed by an internal microphone on both sides and 2 internal speakers to playback sound* (Fig. 2).

4.2 VRiend App

Horus VR built a bespoke app for android devices to support most common standalone VR HMD's like the Samsung Gear VR and the Oculus GO. The VRiend app is user-friendly and the only thing you need to do is to open the app and the livestream will start automatically to a suitable Wi-Fi network. In case of not having a suitable Wi-Fi network you can choose to use a 4G mobile Hotspot.

4.3 VRiend Platform

To set up the low latency and secure connection Horus VR needed to build their own streaming platform. Connection is being established via a signaling server and then the P2P stream will start. The only thing a client needs to do is to turn on the VRiend camera and the VRiend app, connect to a suitable 4G/wifi network and the stream will start.

4.4 Benefits

The breakthrough is Horus VR developed a scalable VR livestreaming platform. This in contradiction to current market solutions VRiend stands out in ease of use and is very affordable (Fig. 3).

Fig. 3 Daughter Emma falls asleep in the hospital with a bedtime story, read by her father, from her own bedroom via a VR livestream

5 Feedback from End Users

VRiend is being used on daily base in over 10 hospitals in The Netherlands. And the first VRiends are flying out to Belgium, Germany, Canada, Japan and France.

The Feedback we got from end users are very positive. A big smile on their faces and very happy to see their class mates again. Or even tears when patients can attend life changing moment like a wedding of a relative or even a funeral. And the healthcare professionals are amazed by the ease of use and what VR livestreaming can do to increase patient's wellbeing for a moment.

The major breakthrough is the outcome that VR experiences where people can identify and even interact with, are more useful and powerful to distract patients and in certain use cases can increase patient's wellbeing.

Horus VR is still waiting for the first results of the research studies. Horus VR is expecting the first outcomes and results end of August 2019.

5.1 Media Awareness

VRiend is being covered in many news items and publications over the last 2 years. With news-items at all major Dutch TV stations including RTL, SBS, NOS. Our VR Wedding was covered frontpage at the best read Dutch newspaper and featured at the 8 O'Clock news. But VRiend made it also to major TV stations in Germany, the UK and even in Romania.

5.2 Nominations & Awards

VRiend won the Dutch 5G innovation award in 2019. A jury, consisting of 5G professionals from KPN, T-Mobile Netherlands, Ericsson and Deloitte, selected VRiend, considering this as the most innovative and promising 5G idea. VRiend was the runner up of the Samsung contest "Best VR idea in NL", ranked in the top 100 of best small companies in NL and finally we were the runner up in Prix Galien Medtech award 2018. Finally, VRiend is being tested in clinical trials and researches all over the world. And we can't wait to share the first results! For a full list of publications, news items we refer to our news archive (Fig. 4).

6 Future Outlook/Roadmap

6.1 VRiend 2.0 Roadmap: Remote Care Tool Set (in Development)

VRiend 1.0 is perfect to communicate via audio and video but it is still a bit limited in the number of use cases. Therefore, Horus VR wants to add new functionality to VRiend make more versatile in different use cases.

Adding functionality doesn't mean a total redesign of the VRiend. Horus VR can add easily new e-health care devices that measure real-time bio signals to the VRiend framework. The only precondition we have is these devices can connect via Wi-Fi, Bluetooth Health services or ANT+ protocols. This means VRiend 2.0 becomes the ultimate remote monitoring toolkit including a live VR audio & video communication-toolset enriched with real-time bio-feedback (Fig. 5).

Fig. 4 Prix Galien Nomination Prof. Marlies P. Schijven MD PhD MHSc

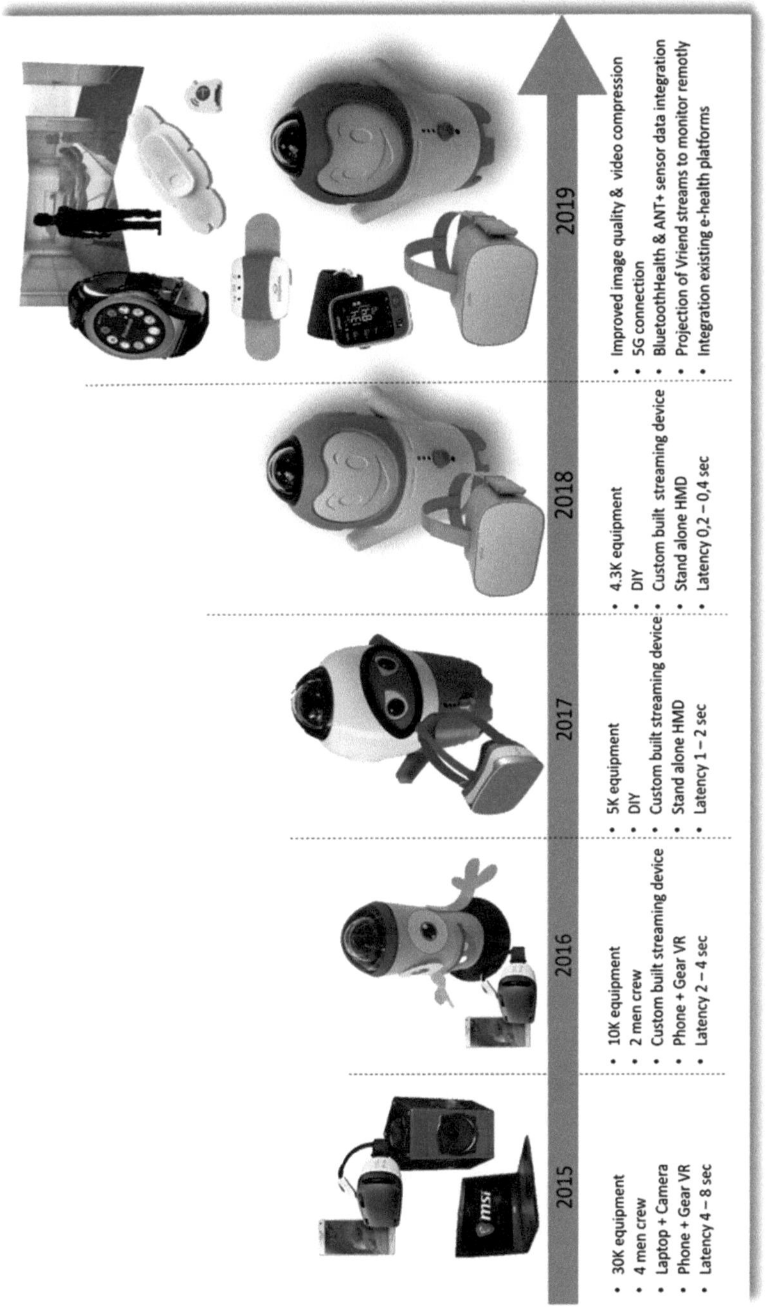

Fig. 5 Roadmap VRiend

7 Conclusion

The major breakthrough is the outcome that VR experiences where people can identify and even interact with, are more useful and powerful to distract patients and in certain use cases can increase patient's wellbeing. Therefore Horus VR developed a scalable VR livestreaming platform. This in contradiction to current market solutions VRiend stands out in ease of use and is very affordable.

Horus VR's purpose, to improve patient's wellbeing by delivering world class infrastructure, has been achieved through the successful implementation of this system in many hospitals and care related organizations.

A very useful addition to the current solution would be the ability to add extra sensor data on top of the VR livestream and getting an even better overview and insight how the patient is feeling. This would open up many more use cases in the cure and care industry.

Acknowledgements Horus VR would like to thank the Horus team for their support in developing a system that allowed the team to realise their ideas in a valuable and meaningful way, using cutting edge technology and innovative methodology. Besides this Horus VR would like to thank the team of the Martini Burn Center in Groningen.

MIX
Papier aus verantwortungsvollen Quellen
Paper from responsible sources
FSC® C105338

If you have any concerns about our products,
you can contact us on
ProductSafety@springernature.com

In case Publisher is established outside the EU,
the EU authorized representative is:
Springer Nature Customer Service Center GmbH
Europaplatz 3, 69115 Heidelberg, Germany

Printed by Libri Plureos GmbH
in Hamburg, Germany